KU-381-155

Open Systems

A BUSINESS STRATEGY FOR THE 1990s

Open Systems

A BUSINESS STRATEGY FOR THE 1990s

Pamela A. Gray

McGRAW-HILL BOOK COMPANY

London · New York · St Louis · San Francisco · Auckland · Bogotá · Caracas
Hamburg · Lisbon · Madrid · Mexico · Milan · Montreal · New Delhi
Panama · Paris · San Juan · São Paulo · Singapore · Sydney · Tokyo · Toronto

Published by

McGRAW-HILL Book Company (UK) Limited
SHOPPENHANGERS ROAD · MAIDENHEAD · BERKSHIRE · ENGLAND
TEL 0628-23432 FAX 0628-770224

British Library Cataloguing in Publication Data
Gray, Pamela A.
Open Systems: a business strategy for the 1990s.
1. Computer systems. Communication networks.
International standards. Open Systems Interconnection
I. Title
004.62

ISBN 0-07-707244-8

Library of Congress Cataloging-in-Publication Data
Gray, Pamela A. (Pamela Anne),
Open Systems: a business strategy for the 1990s/Pamela A. Gray.
p. cm.
Includes bibliographical references (p.) and index.
ISBN 0-07-707244-8
1. Electronic data interchange. I. Title.
HF5548.33.G73 1991.
658'.05—dc20 90-13273
CIP

12345RC 94321

Typeset by Burgess & Son (Abingdon) Ltd.
Printed and bound in Great Britain by Clays Ltd, St Ives plc

To the constant in an ever-changing world:
my family

Contents

1 Introduction and overview 1

2 Standards in the computer industry—who needs them? 8

3 The standards needed 25

4 The making of standards 44

5 Standards for portability and scalability 65

6 Standards for interconnection 96

7 The role of X/Open 121

8 Users of open systems 142

Appendices

1 Data handling 167

2 The user interface 187

3 Internationalization 213

4 Security 227

5 System administration 238

Glossary 250

Sources for further information 254

Index 257

Foreword

It is unusual to find a book on computing which can be read profitably by technical and non-technical persons alike. This is such a book. It shows technical people the non-technical implications of 'open-systems' computing, and introduces non-technical people to the world of international standards, the driving force behind the Open-Systems movement.

'Open systems' is a term which is much used, but rarely understood. This book defines 'open systems' unambiguously and authoritatively; it will undoubtedly become a standard point of reference for the future.

The author, Dr Pamela Gray, is a leading advocate of open systems. When most computer professionals thought of UNIX as a technical product, she realized that UNIX-based systems and applications would revolutionize the computer marketplace. This vision has become reality. POSIX, the international standard based on UNIX, is now well-established as the standard 'open' operating system technology for multi-user, multi-tasking systems.

But even more importantly, the concept of 'open systems' has led to the growth of an open market for computers. All significant hardware manufacturers now support the open-systems movement. Through their participation in X/Open, companies which normally compete with each other work together to create and support industry-wide standards.

The emergence of standards and industry-wide support for the standards movement has, in turn, led to the expansion of this market. UNIX-based systems now represent about 15 per cent of the world computer market, and the market share is growing.

Dr Gray's contribution to the creation of this exciting new market has been recognized by both the computer industry and computer users. She has twice been elected president of Uniforum, the US-based users' organization which initiated much of the early work on standards. She is now chief executive of Marosi Ltd, which she founded in 1989 to provide open-systems information and services to user companies.

For end-users, open systems mean choices and challenges. As companies become less dependent on proprietary systems, they need to learn how to function within a different and freer marketplace.

This means that they must be able to understand the way in which international standards emerge. They must be able to track those standards, and make decisions about which standards are relevant to their particular needs. Finally, they must be able to evaluate hardware and software in the light of those standards.

This book is a good place for them to start. It should be read by any technical manager, and any non-technical person who may be involved in the acquisition and implementation of open-systems products, hardware and software. Above all, it should be read by anyone who wishes to be well-informed about the most dynamic marketplace within the modern computer industry.

Geoff Morris
President and Chief Executive
X/Open Co. Ltd

Preface

This book is a result of my frustration over the last few years both with the seeming inability of the computer industry to react fast to the practical needs of its customers, and to the scarcity of materials which explain to those who need to make strategic decisions in this area what the relevant issues are and why they are so important.

It is my contention that an information systems strategy based on 'open systems' is the only logical way forward in the 1990s for most organizations, and that anyone who understands both its philosophy and the business and economic issues behind it will agree with me. So I have attempted to explain here the extent of the open-systems movement, in a way that can be understood by anyone who wants or needs to do so.

I have assumed no prior knowledge of the subject and have tried to keep jargon to a minimum—not an easy task in this industry. Since the world of standards and open systems abounds with acronyms and abbreviations, these are listed in the Glossary. The first time any one of these is used, it is spelt out in full; thereafter, it is referred to by its abbreviation. I apologize in advance for the fact that there are so many of these, and that I was unable to find a way to avoid their use.

The book is organized into two parts. The first part (Chapters 1–8) covers the subject overall, making mention of a number of issues that are of great importance at the present time. The second part (Appendices 1–5) covers these particular issues in greater (technical) detail for those who may be especially interested. I would like to thank Peter Judge, author of 'Open Systems: The Basic Guide to OSI and its Implementation', published by *Computer Weekly*, for allowing me to use some of his material within Chapter 6; X/OPEN Co. Ltd, for allowing me to reproduce some of the user stories contained in its monthly newsletter, *Open Comments*, within Chapter 8; and to my long-term colleague and friend Dominic Dunlop, for providing me with much of the material contained in the Appendices.

I have included a list of sources of further information which is necessarily limited. However, some of the organizations mentioned will be able to supply a great deal more material in this area, upon request.

This book represents my own views on open systems; these are not necessarily shared by all participants within the computer industry. I have at all times tried to be fair, withstanding the temptation either to criticize or support the work of any particular product vendors.

I would like to thank all the people who have contributed to this work, both directly and indirectly. Many of these must go unacknowledged, being the many people in organizations around the world who have heard me speak on this subject, and have volunteered their comments. Much of what I have come to believe is based on their input.

Particular thanks are due to Stuart Strolin, who read and edited various drafts; Brian Boyle, Ellen Hubbard, Guido Van Herbigen who reviewed the final draft; Amanda Stuart, who provided administrative support and encouragement; and Janet Wass, who patiently coordinated the final production process.

X/Open Co. Ltd, pioneers of much of the work in open-system standards, provided positive support by funding a parallel project on standards activities while I was writing this book. Some of the necessary information on standards was obtained under that project, and I would like to express my thanks for this. It is my opinion that X/Open represents the greatest force for the eventual implementation of the many standards that the computer users need, and I am proud to have been associated with it.

When seeking a publisher, McGraw-Hill was my first choice. I would like to thank the people involved for their confidence, as demonstrated by the rapid agreement to proceed, and for the high quality of the professional services throughout the publishing process.

CHAPTER 1 Introduction and overview

> There are two things that most senior executives know about information technology. The first is that it costs too much. And the second is that it never works.
> *Management Today*, April 1989, in a review of *Management Strategies for Information Technology* (1989) by Michael Earl, Director of the Oxford Institute of Information Management.

This statement probably sounds cynical to anyone who does not work at the user end of the computer industry. But unfortunately, to those who do, it is a succinct reminder of the real state of an industry that promises so much and yet often delivers so little.

The objective of this book is to define, in language which is general enough for any interested reader to understand, some of the technological and political issues that lie behind the problems which many computer users experience. It aims to demonstrate that there are potential solutions, but because varying conflicting forces exist, these may be difficult to attain. These solutions will be shown to require cooperation on many fronts from organizations that are more used to competing with one another.

Rather than taking a technology-driven view, we will investigate the issues through a 'top-down' approach, looking first at the general problems and applying common-sense to our search for solutions. Only when the overall method for solution becomes clear will we look into the technical methods available for implementation.

In this chapter, we will take an overview of the basic issues which lie behind our opening quote. We will see that there is a clear and logical path to take in order to achieve improvements, but how we are all to get on to it is not so clear. That will be the subject of the later chapters.

The worldwide problem

Over the last few years, there have been reports from many countries of problems with computer installations. These cover many areas but can briefly be characterized under three main headings:

- computer project costs and time *overruns*, sometimes by massive factors;
- application *backlogs*, where the time for the data processing department to write software required for particular applications has become unacceptably long;
- projections of increasing personnel *skill shortages* over the next few years, with the result that problems are anticipated to get increasingly worse.

The results of one such study, undertaken recently in the US, are summarized in Figure 1.1.

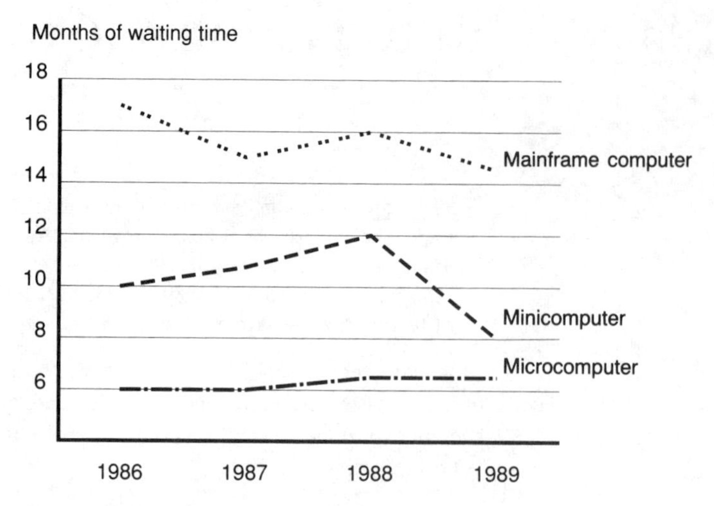

*Figure 1 .1 The application software backlog in US (*Software Magazine*, March 1989)*

Here it can be seen that the average delay time before work is started to develop a required application on a company's mainframe computer is approximately 15 months. Even on the ubiquitous microcomputer, it is more than 6 months. (The reduction in delay time for minicomputer application software is more likely to be a drop in the requirement for such software rather than an increase in the efficiency of development.)

The frustration caused in the user community by such application development backlogs is in part the reason for the proliferation of personal computers within organizations. Tired of waiting for the central data processing (DP) or management information services (MIS) department to produce the tools they require to do their jobs more efficiently, many users have taken the law into their own hands and have developed their own applications. Although this has solved one problem—decreasing the development delay time—it has produced another. The lack of central control has led to a distribution of data and information within many organizations, with this now contained within and on a number of disparate systems scattered throughout the company. It is a major problem to develop the means to bring this hardware and software back into one fully integrated information system to serve the needs of the organization as a whole. We will return to this later.

The problems of project overruns, development delays and skill shortages are not confined to any one country or industry. This is illustrated by many recent investigations undertaken in Europe as well as in the US. One of these, conducted by the Amadahl Executive Institute in 1989, surveyed 300 European companies across several industry segments, looking specifically at their usage of computer technologies. The study concluded that over two-thirds of the companies were failing to implement information systems to the level of success which is thought necessary in order to keep pace with future business needs.

Similar results have been found in detailed studies within individual countries. For example, in 1984, a UK Department of Trade and Industry report into the use of information technology by 235 UK companies estimated that British companies wasted £800 m a year on inappropriate computer-based projects. No company was achieving more than 80 per cent of the potential benefits and average benefits were put closer to 50 per cent. Since then, industry commentators believe the problem has, if anything, worsened.

Quantitative information to substantiate the costs of wastage contained within practical implementations of computer technology is very difficult to come by. Most companies have little or no idea of the *real costs* of their information-technology installations, having few mechanisms in place to track many of the indirect or hidden costs.

Many of these are contained in such nebulous matters as the cost of transferring 'know-how'—training and retraining as new people are hired or moved around within an organization, or as new systems are installed—and the costs to the business of inefficiencies in the systems used. Some studies have taken place in the US attempting to track the real 'costs of ownership' of purchased systems, but to date the methodology for tracking total costs is immature and not used by many organizations.

Little information is readily available to date on the situation prevailing in Japan, Taiwan and other countries in the Far East. Anecdotal stories (and history) lead one to believe that Japan in particular may be managing the problems better than the rest of the world. However, they undoubtedly will also face future skills shortages, which are predicted to become a worldwide problem. System suppliers from these and other countries will also face problems selling products into other countries, if they do not cooperate with suppliers on a worldwide basis to develop products that will help resolve the problems that are at the root of many of today's difficulties.

Invisible costs

IT installations involve huge investments—visible in the form of hardware and usually obvious in the costs of software and software development. But they also involve a huge investment in the accumulation of know-how, expertise and the 'feeling of comfort' contained in people when they become proficient in the use of the system. A quantitive study of annual DP expenditures in the US by the Gartner Group demonstrates this (see Figure 1.2).

As the illustration shows, direct hardware and software costs in this study only accounted for 32 per cent of the total expenditure. The remaining 68 per cent covered all the add-on costs associated with the system—communications, maintenance, training, support, and so on.

Many of these additional costs are 'soft' costs buried in such people-related matters as:

- the training and retraining that takes place continually as personnel are

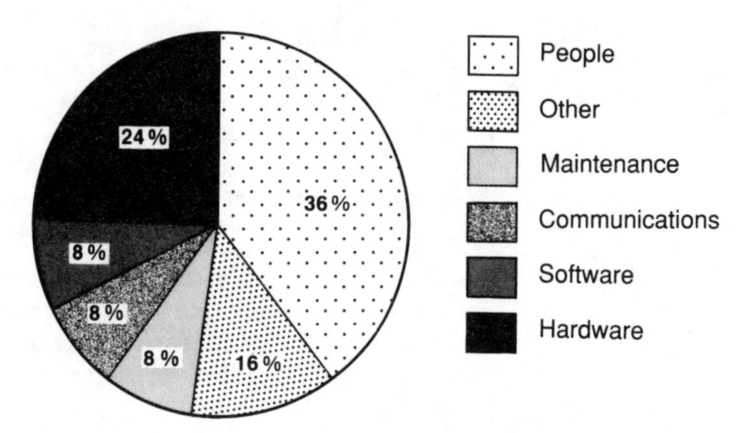

Figure 1.2 Annual DP expenditure in major US companies

hired-in or moved around within an organization or as the systems are updated and changed;

- the less than optimal results that are obtained by the business when the systems are inefficient;
- the disruption to the business that results when the system breaks down or is changed.

Others relate to installation, maintenance and wiring costs, for example:

- wiring, or re-wiring, of the building;
- air-conditioning, for the larger systems;
- increased telephone usage via modems for communications between people and/or computers;
- power and computer supplies.

Many of these additional costs—for example, the cost to the business when the computer system is inoperable ('down')—are difficult or impossible to quantify, as are many of the people-related costs.

Costs of resistance to change

The investment made in the accumulation of know-how within the heads of its personnel, both through formal training and by experience 'on-the-job', constitutes a large and almost invisible asset for an organization. But unlike the situation that prevails for most assets in today's businesses, it is difficult to find an organization that appears to manage it efficiently, perhaps because it does not yet appear on the balance sheet. However, it represents an investment that no one within an organization would wish to see discarded unnecessarily.

Unfortunately, the invisible asset of accumulated know-how sometimes makes its presence felt in a negative way. The people in whose heads the know-how resides sometimes show an automatic resistance to those changes that threaten the feeling of comfort they have developed. Often, the resistance is not to change itself but to the 'discomfort' of having to relearn a new system. Such resistance can itself increase the hidden costs of a new system.

The general solution

Today, it is clear that efficient use of information technology for maximum exploitation of a company's resources in increasingly competitive environments is a crucial strategic issue for many organizations. Yet in spite of this, the facts seem to indicate that the IT industry has been slow to understand, and is therefore not properly serving, the most urgent needs of many of its largest clients.

For those who work within the computer industry, it is easy to be misled into thinking that such statements are exaggerated. Too often they deal directly only with organizations who have sophisticated computer installations, and have the resources (and the desire to commit them) to track the technologies and to interpret and implement them for their own specific needs.

But such users represent only the tip of the iceberg. The bulk of potential and actual corporate computer users make up the rest. The majority of these struggle against cost overruns and doubtful functionality in an increasingly competitive world. Gaining a competitive edge through the use of IT means keeping up to date but advances in technology proceed so fast that this often seems an impossible task.

In fact, it sometimes appears that technology is advancing faster than people's ability to assimilate it. It is certainly advancing faster than the availability of skilled personnel in the population at large, causing the skill shortage problems to exacerbate. Thus the computer industry needs to find ways to allow both developers and potential users the ability to utilize new technology faster than they are able to at present, while at the same time insulating them as much as possible from the detail of the changes within. In other words, IT should be made easy to use, while at the same time designed in such a way as to allow innovation and competitiveness among suppliers to proceed.

We contend that the only way to do this is through the adoption of international, widely used, consensus standards for IT systems, supported by the entire computer industry, and set in such a way as to reflect the real needs and priorities of the marketplace.

It is our objective to support this contention by analysing the needs of various classes of users of IT, including the providers of many of its components, and indicating the myriad of circumstances where these can best, and sometimes only, be addressed by strategies based on compatible, standard computer systems and associated technologies. Within the computer industry today, such compatible systems are described as open systems.

Open systems

Although it is generally agreed that open systems mean in some sense, compatible systems, much confusion exists about the actual definition of the term.

This is partly because compatibility is required in many different ways in order to satisfy many different requirements.

For example, some take open systems to mean 'standards for the computer industry', defined by a process that is 'open' in the sense of its being possible for any interested party to participate, with no advantage given to any one participant. This could be considered the 'political' definition of open systems.

Others consider open systems to refer to the ability for computers to communicate easily, one with another, in a similar way to that in which people around the world can communicate with each other over the telephone. Thus one finds organizations such as the Corporation for Open Systems, which concentrates almost entirely on defining standards for communication. This definition of open systems is the most commonly accepted one in the European computer market.

What happened to the portability definition of 'open'—most commonly accepted in the US?

Yet a third interpretation is made by those who consider open systems to provide complete standardization of the computers themselves. When this is achieved, a software package that works on one computer will work on all others of the same standard class. This already applies today to computers that are totally compatible with the IBM PC.

Interestingly, many of the people who are active in the open-systems movement do not consider the IBM PC and compatible computers to be 'open'. This is because the technology has been defined and controlled by a single company (IBM) and has been copied by almost everyone else. In the 'political' sense, and in the communications sense, these are *not* open systems. In the sense of providing compatibility across machines from different manufacturers so that software designed for one will run on all, they are.

One reason for confusion in the definition of open systems is that the computer industry too often focuses on the detail of the technological means for achieving openness and not on the desired end itself. When the economic and business issues that are driving users to open systems are kept in mind, it becomes easier to understand whether or not a system or product fits with an open strategy.

Throughout this book, we will consider 'open systems' to be a catch-all phrase for the multitude of technical standards that will have to be set to produce the end that the users require. Thus it will encompass all three concepts described above, although our focus will be principally on standards for portability of software, particularly for multi-user systems, and on the communications facilities required for them.

Summary: introduction and overview

In the following chapters, we will define the technical standards required for open systems and analyse their components, looking at each in terms of user requirements. We will see where the industry currently is in its attempts to define the standards, and where the user community is in its uptake of them. The organizations which are key to the open-systems movement will be described and their contributions analysed. Finally, we will give examples of user organizations that have adopted open-systems strategies in various forms and will offer advice to organizations wishing to use the open revolution in the computer industry to their own company's advantage.

It will be shown that forces for standardization across the computer industry are principally economic. When coupled with the notion of 'making information technology easier', the forces will be seen to be logical, rational and produced from user needs. For this last reason, open systems based on standards will ultimately prevail.

CHAPTER 2

Standards in the computer industry—who needs them?

Why standards?

Until recently, the computer industry has been driven by its technology rather than by the requirements of its market. It has few of the customer surveys of needs and requirements, priorities and problems that guide the more obvious consumer industries. Instead, its market research is primarily 'after-the-event' data, which is then extrapolated crudely to give forward forecasts of expected shipments of hardware and software products.

Marketing strategies of computer suppliers have too often been based on the principle of: 'Here's a great product—now where shall we sell it?'. Until recently, many computer system supplier companies perceived their competitive edge as achievable through differentiated, proprietary (available only from them) technologies. As a consequence, there are many technologically superb products in the computer marketplace but few people able to use them efficiently.

Applications software written for a computer from one particular manufacturer has previously been specific to it and not easily moved to run on another manufacturer's machine. And such fundamentally different machines have not been designed to communicate easily, one to another.

The incompatibilities between computer systems lead to a dramatically different situation from that which prevails, for example, in the hi-fi industry. There, all manufacturers' systems can play the same discs, cassettes and CDs. Competitive differentiation is on price, performance, functionality and appearance. In other words, the hi-fi industry supports common standards for music and media while allowing plenty of scope for innovation and competitiveness among suppliers.

Standards which allow innovation within the system design without destroying the basic compatibility across systems from different manufacturers give the customer freedom of choice of system. In the hi-fi industry, the choice is usually based on budget, with the customer having confidence that any system purchased can accept standard musical media. Investments made prior to the purchase of the new system are protected, in the sense that the new system is able to play the old music. Of course, the media variations can be annoying and expensive. It would be highly attractive, for example, if one could convert from older technologies, such as LPs, to newer, such as CDs, without having to purchase all one's favourite music over again on the new media.

In the video industry, it appeared for some time that two standard technologies, Beta and VHS, might survive for videotape formats. In some ways, Beta was considered technologically superior, while VHS was available from more video-recorder suppliers. But the forces that led to the supremacy of VHS worldwide were *not* primarily technological. Suppliers of video-cassettes could not, or would not, accept the costs of carrying double inventory, nor would purchasers accept that the cassettes they had purchased (or borrowed) would not run on any system. The economic forces—*the forces for simplicity*—eventually drove a single standard, VHS.

Proprietary lock-in

Until the advent of the microprocessor and the computer revolution that followed, computer manufacturers had managed quite successfully to lock their customers into a single supplier purchasing policy by providing systems that would only run software written specifically for them.

Once software applications had been developed, it was often impossible for the users to switch hardware systems suppliers without having to throw away huge investments made in software development, together with that made in the accumulation of specific technical know-how.

Apart from the obvious risk of non-competitive pricing, the *proprietary-systems* approach has led to increased costs and inefficiencies both in development and throughout the supply channels for both software and hardware products worldwide. These additional costs must always be reflected in the eventual costs paid by the users for those products.

Market forces for standardization

In order to understand the forces that are operating in the computer industry as a whole, and in particular, to see where the forces for increased compatibility and standardization are, it is useful to look at each of the elements that make up the computer industry market (see Figure 2.1). As we shall see, analysis shows a large number of factors, many common to the various market segments, which are driving towards the adoption of common standards across the industry.

The issues for packaged software developers

Software developers today are faced with hugely escalating development costs. Software products are becoming increasingly complex. Control of the development process is now so difficult that it is common for an announced software product to be delayed months, or even years, past its expected release date.

Increased sophistication of user expectations, coupled with rapidly evolving technologies, means that software products are taking longer to develop and lasting a shorter time in the marketplace than ever before. Recovery of investment is increasingly difficult and can only be made by selling the software on the largest possible base of hardware systems; hence, the large number of developers that focus on the IBM PC and compatible systems market.

No software author today can afford to target software products exclusively

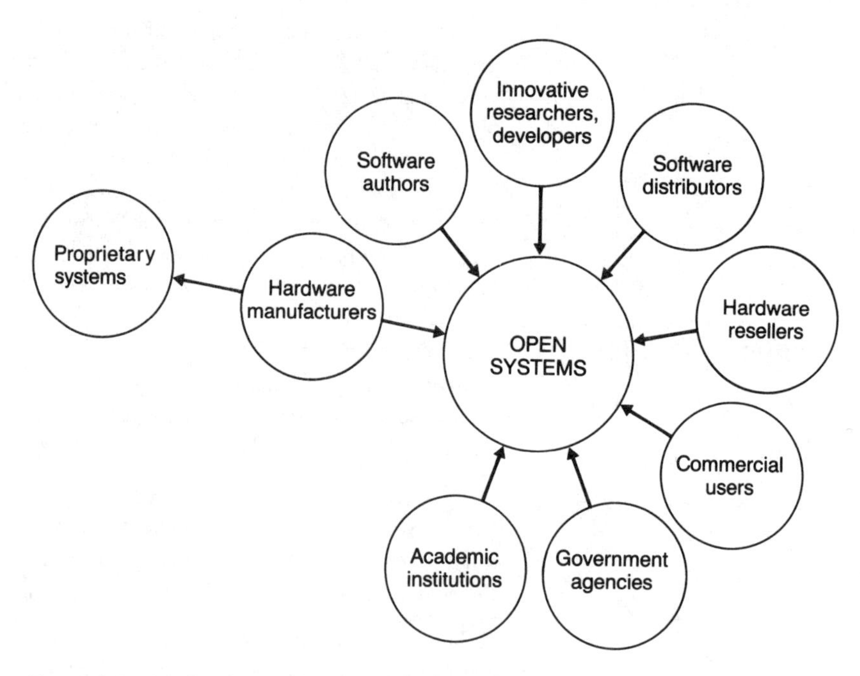

Figure 2.1 Market forces for standard (open) systems

to proprietary environments unless sales can be guaranteed to be high. Few computer manufacturers can today offer such guarantees.

A presentation by Bill Gates, Chairman of Microsoft Inc., at the US Uniforum Conference of 1985 made this point very clearly. Gates claimed that the number of units of software that needed to be sold in order to support software development and subsequent distribution sales costs was at least 400 000 units on *totally compatible systems*.

The costs of 'porting' software

Some software companies have nevertheless become enormously successful through offering their products on a number of proprietary hardware platforms supplied by the large computer manufacturers. A notable example is the US database supplier, Oracle Inc.

Very early in its history, in order to provide sufficient volume for its ambitious growth plans, Oracle produced its database products for almost all of the major and popular proprietary systems. More recently, Oracle has expanded its range to include the non-proprietary (standard) open systems. This far-sighted (or opportunistic) strategy allowed commercial organizations to develop a corporate database strategy, based on Oracle products, that was independent of the choice of supplier of the hardware platforms.

By standardizing on Oracle database software products, each department in an organization could choose whichever computer system was appropriate to its own price/performance needs while retaining compatibility at the applications software level across the corporation. But costs to Oracle of this program have been high. The work involved in moving ('porting') a software product from one proprietary system to another often takes many man-years and costs millions of dollars.

If, however, two different systems manufacturers adhere to a common standard technology, porting time can be dramatically reduced. When there is source code compatibility (standard systems but using different microprocessors), porting takes a matter of days, with costs in thousands of dollars. In the case of binary portability (standard systems using the same microprocessors, as in the IBM personal computer and its clones), there is no cost of porting; software runs on all such compatible systems automatically.

Porting is not a 'once-only' job. When there is a new version of the hardware system, and its operating software, or a new revised version of the application software product, or when the software product is translated into a foreign language, all the previous porting work must usually be repeated.

Computer manufacturers, who often supply software products directly to their customers, sometimes seem to have absorbed these porting costs themselves. Within their software distribution contracts with the software authors, porting charges may not be obvious but they will be there, embedded in the software prices. And whether visible or not, eventually the additional costs will be passed on to the final consumers and users.

The cost of producing a different software version for each proprietary system, combined with the number of software units that need to be shipped overall for profitability, means that software costs are unnecessarily high. Clearly, by adopting system standards, the computer industry could start to supply to the computer marketplace the quality, value-for-money and range of choice of product that is enjoyed, for example, in the hi-fi industry. In fact, this has already happened for software products in the PC industry, where systems are totally compatible with the IBM standard.

In the multi-user market today, there is a 'standard' system (based on the UNIX operating system) available on computer systems supplied by almost every manufacturer in the world.

However, there is not as yet complete compatibility between most manufacturers' systems as shipped. In fact, there are currently more than a hundred variations of this so-called standard system that must be supported by any software author wishing to have reasonable market coverage. Thus, although the costs of porting between any two of these, almost standard, environments are acceptable, when multiplied by more than one hundred, they are certainly not.

In such a highly fragmented situation, there are again huge costs which are accumulated and passed along the software distribution and support lines. Again, of course, these are ultimately paid for by the users.

Indirect porting costs

It is not only the direct costs of the porting work which must be considered. There is also the cost to the developer of obtaining, maintaining and running the myriad of computers that are needed—at least one of each kind and flavour—in order to provide technical support for any particular version of the software in the event that problems arise.

There is the cost of space for housing these machines, and the costs for the air-conditioning, power and people to look after them. A further irritant to the market is introduced by the time delays caused by the long cycles in the porting work. Releases of new products cannot appear simultaneously on all machines. Consequently, application software versions on different manufacturers' computer systems at customers' sites then become out of phase. Supporting and updating software products already in the field can become an administrative nightmare.

Conclusion

Many of the problems experienced today by software authors and developers could be dramatically reduced if the computer industry were to move to the supply of standard hardware systems supported by the major manufacturers.

Since the porting work on software needs to be done anew every time there is a new version of the software or of the system on which it must run, or a translation of the software into another (foreign) language, the advantages of standard, open computer systems—open in the portability sense—are overwhelming for software authors and developers.

The issues for software distributors

Today, most software is supplied to the eventual users through indirect channels, rather than directly from the software authors themselves. These channels may be *software distributors*, value-added resellers (VARs) or even the hardware manufacturers.

None of the links in the 'software supply chain' wishes to absorb costs for porting and re-porting of software to various machines. All of them would like to keep their inventories to a minimum while offering to their users a high level of technical support on the software products themselves at the lowest possible costs.

Inventory and support

Software distributors and resellers face the additional problem of inventory control. The greater the number of different systems that exist, the greater is the number of variants of the software product that the distributor must hold in stock. The higher the inventory required, the more the problems of inventory management are compounded.

In order to offer good technical support on the application products to be supplied, it is usually necessary to have available a computer which matches that which the particular client is using. It is also necessary to have expertise available on that particular system in order to ascertain whether the problem the user is experiencing is with the application software or with the system itself.

At present, in order to maintain sufficient sales volumes, it is necessary for software distributors to supply and support software products on many incompatible systems. This has led to increasing costs for inventory management and technical support. Coupled with the low margins at which most distributors operate, many distributors are finding it difficult to stay in a straight distribution business.

Software distributors supplying exclusively to the IBM PC and compatibles market were exceptions to this until recently, when market trends have started to force them to supply software products for both the current (PC-DOS) and emerging future (OS/2 and/or UNIX) standard systems.

Conclusion

If there were to be a much smaller number of standard platforms on which to supply applications software, distributors could minimize their inventory problems, while maximizing their ability to offer good technical support on the applications themselves.

They would require fewer computer systems for the support process and fewer system level experts in today's proprietary environments. Thus, standard (open) systems—open in the portability sense—provide an attractive scenario for software distributors.

The issues for hardware resellers

As with software, it is also the case today that many computer systems are delivered to the users by a complicated distribution chain involving a range of specialized *computer hardware resellers*, and not by the computer manufacturer directly. These are referred to as systems houses and value-added resellers (VARs), and include networks of franchisees, as well as numerous distributors and small dealers.

These resellers, in general, package up hardware with specialized applications software targeted for sale to particular specialized vertical markets—for example, insurance brokerage or hotel reservations. The applications software, while adapted and customized for these specialized markets, is often built from packaged products such as standard databases, program generators, fourth-generation languages (4GLs) and communication modules.

The need for choice

These resellers do not want to absorb any additional costs that might be built into the software prices by the authors and the software distributors from whom they buy. In addition, they would ideally like to be able to offer a choice of price and performance on computer hardware while retaining the ability to supply and easily support *exactly the same* software solution.

Even when the customer is dedicated to a particular hardware supplier, a certain level of comfort and control is provided when both customer and supplier know that, should conditions warrant, it is not prohibitively expensive for the customer to switch suppliers. Standard systems which are open—in the portability sense—would accomplish this.

Interconnection requirements

In order to offer suitable price/performance options, systems resellers usually sell their application software products on a variety of computer hardware systems. Because they sell these into customer sites that usually have other systems already installed, it is often necessary for them to integrate disparate machines into an integrated whole, using communications technologies. The more compatible the systems are, and the more standard their communication methods, the easier the integration work is likely to be. For this reason, standard open systems to the system resellers usually means standards for communications also.

Conclusion

For the system resellers, standard open systems could be expected to keep costs down, allow a choice of systems supplier without the need to adopt the software, and allow the provision of consistent technical support across different manufacturers' machines.

'Open' for the reseller means not only portability of software, but also the ability for systems to communicate and integrate easily with each other, as well as with systems that already exist at customer sites.

When compatible systems supporting both portability and interconnection standards are available from a number of suppliers, with a range of price/performance options, the current situation would be dramatically improved.

The issues for computer users

For the users themselves, whose real needs should always be the top priority, there are clear potential advantages to having standard open systems available from a choice of system suppliers.

First, competition has always led to improvements in product price/performance ratios. Users who have suffered from the excessive pricing sometimes produced by proprietary lock-in have learnt that a level of freedom of choice of supplier is vital, provided that it can be obtained without sacrificing the functionality of the products.

In certain *government* procurements, particularly those involving defence, alternative sources of supply are often a requirement. Usually this is applied to the components of the system, rather than the system as a whole. The requirement is then for compatibility in form, fit and function of the component parts. As we shall see later, government procurement policies have driven much of the early work on standardization.

Universities and training establishments are often at the forefront of technological development. For them, the ability to support rapid changes in technology, while at the same time addressing the breadth of technology, is a significant challenge. This could be achieved with open standard products from a variety of manufacturers without detracting from the opportunities for innovative development. The widest possible future employment options would then be open to graduates, who will have gained experience pertinent to the open systems marketplace.

In general, proprietary systems impose heavy overheads in support, training and development on user sites. People whose know-how is specific to proprietary systems cannot easily be moved within or across organizations that use different systems. Standard open systems lead to portability of people, as well as of software products, and increase the applicability of experience to the marketplace in general.

Skill shortages and people portability

Many organizations around the world are finding that the constraints to their productivity and growth are increasingly caused by a shortage of skilled people. As an illustration of the problem, Figure 2.2 shows the results of a study undertaken in the UK in 1987. There, the divergence between the demand for analyst/programmers and the potential supply is expected to increase dramatically over the next few years. Studies in other countries show similar forecasts. Although technology itself can provide many ways to increase efficiency in theory, in practice its use is limited if the technology is too difficult to use or there are too few people able to use it.

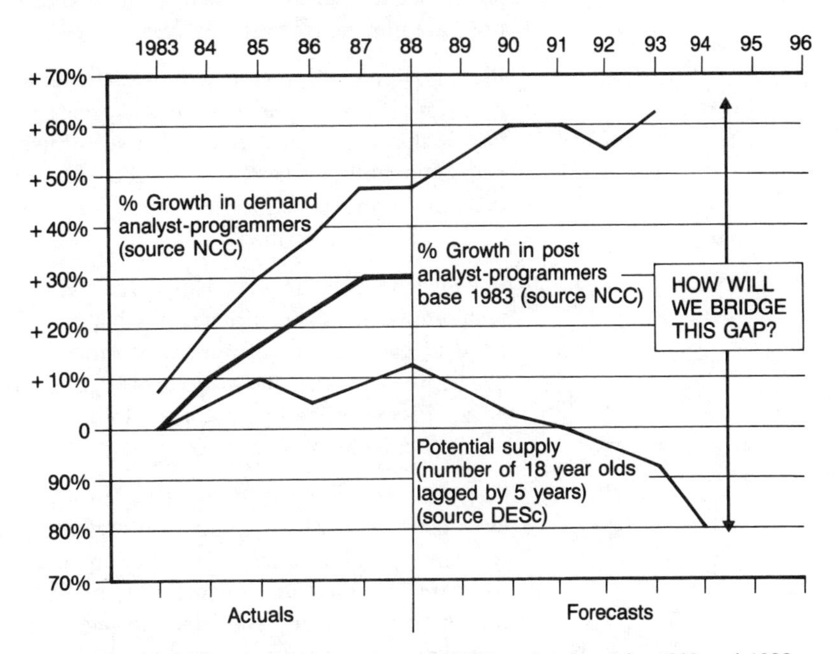

Figure 2.2 Predicted UK IT skill shortages

Across the IT industry there is an urgent need to solve the problems created by the skill shortages. This can be done either by making the technology more easily usable, so that more people can take advantage of the benefits it offers, or by increasing the pool of skilled people through training, or both.

Worldwide, we must find ways to increase the pool of people able to make best use of available information technology products. We need to reduce the learning times of people in new positions and make it easier to move people within and between organizations without the need for constant training and retraining.

A move to standard open systems would reduce this need dramatically, demonstrating an additional characteristic of the open environment—'*people portability*'—where systems are designed in such a way that people can carry useful know-how and directly applicable experience with them when they move.

To illustrate this characteristic: when a person learns to drive a car of any make, we take for granted that he or she can then drive any car from any manufacturer anywhere in the world. This is the case even though there are many minor differences between model designs, and some major differences in local conventions, such as which side of the road the traffic is on. It is this kind of *skills portability* that we will obtain with the adoption of open-systems standards.

Government and commercial users

Although government purchasers worldwide are moving rapidly to system procurement strategies based on international open standards, commercial users are moving more slowly. Why is this?

First, government procurements are almost always sensitive to the requirement for alternative suppliers. And, because government pay is usually lower than in the commercial sector, the issue of minimizing training and achieving people portability is easily recognized as important. In the commercial sector, the long-term economic advantages are clear only when the issues are explained and understood.

Even when this is the case, many large commercial users are heavily involved with proprietary systems, many of which are old. These often carry with them large investments in software which must be maintained for their business needs. In such cases, the manufacturer of the proprietary systems has little or no interest in promoting a change to open systems, since this may well let competitors into the site.

The problem for many users is not in understanding that a move to open systems is the correct one for their organization. Rather it is in defining and implementing the path by which they are to get there. The requirement is then for a *strategy for change* and for the products with which to implement the migration path from proprietary to open. Such a strategy must clearly be implemented without a negative impact on the business and without demotivating the experts whose heads currently contain the know-how required to manage the proprietary systems.

The *management of change* is a very difficult problem, and one to which we shall return later.

Investors

Much of the creativity and innovation that led to the brilliant products which abound today within the information-technology industry originated in entrepreneurial companies funded in their early stages by venture capital and other high risk investors.

Retaining innovation

Either consciously or unconsciously, the investment community is recognizing the advantages of volume that could come from products targeted to standard environments. As a result, more and more advanced development in emerging technologies is based on standards.

Over 80 per cent of computer-related investment from venture capital companies in the US in recent years has involved the use of emerging standards for portability (specifically the UNIX operating system), and open-systems networking standards. Many of the major computer hardware manufacturers now claim to be spending at least 50 per cent of their research and development budgets on the development of products that meet internationally defined standards. If this level of investment is sustained, users can expect to be able to buy high quality innovative products based on standards without having to pay the penalty of proprietary lock-in for the future.

Conclusion

The overall requirement for innovators and their financial backers is for standards that allow portability of software, know-how and people at one level within the system, while allowing maximum scope for innovation and inventiveness beneath. This has been achieved in the automobile and hi-fi industries, and is essential—and clearly attainable—for the computer industry. In addition, systems must communicate easily with one another, both with the systems inherited from the past and with the systems likely to arrive in the future.

The use of standard technologies will lead to higher volumes for the target markets, and therefore increase the probability of realizing value from the investments made.

The issues for hardware manufacturers

Despite the increasing level of their commitment to, and investment in, products based on standards, the major hardware manufacturers find themselves in a dilemma. Used to the high margins that come from their proprietary product lines, it will be difficult for some of them to maintain current profitability within the potentially vast, but highly competitive, market of open systems.

Many manufacturers have made considerable investments in their own technologies over the years, as indeed have many of their customers. Managing the migration from proprietary to open systems while at the same time retaining profitability and perhaps even growing the business, is proving very difficult for both customers and suppliers.

As technology becomes even more sophisticated, and skill shortages worsen, even the largest of the computer manufacturers is finding that it can no longer 'go it alone'. More and more system components are bought in from other companies, or jointly developed, and more and more industry partnerships formed. Many smaller companies have been unable to survive under the heavy load created by development of proprietary technologies in a marketplace that is increasingly demanding openness.

The use of standards provides the manufacturers with a common base on which to do their, often now shared, development, and on which their personnel, both new and old, can be trained. This will produce big cost savings eventually in the industry as a whole, and allow its brightest people to concentrate on applications for the future, rather than re-inventing wheels for today.

Conclusion

For the hardware manufacturers, the market forces for open technologies are now so strong that those manufacturers who do not respond actively, and fast, are likely to be left behind. Both suppliers, and their customers, need a competitive edge and will need to move to standard technologies to achieve or maintain this in the future.

To underline the importance of this issue for hardware manufacturers, the US Federal purchasing budget, valued at more than $2 Bn a year, now requires the use of industry standards in more than 80 per cent of its procurements. Governments worldwide are adopting similar policies. No manufacturer is so large that it can afford to ignore such a market.

The case for standards

Having now covered, if only briefly, the issues that are important and relevant to each of the computer industry segments that are driving the open-systems movement, as illustrated in Figure 2.1, the common requirements should be clear. Many of the problems could be solved by technologies designed to allow portability of applications software, portability of people and their skills, and ease of communications between systems from various manufacturers. Once achieved, such open systems would allow freedom of choice in purchasing and focus of creative energy on the real-world application problems.

We leave this section on a philosophical note—one that may give some comfort to those computer manufacturers and commercial users who are finding it hard to adjust to the open-systems revolution.

A note on freedom

Freedom of choice is an absolute necessity for people to have in most circumstances in the world. But freedom is a strange thing. When people do not have it, they will fight to the death to attain it. When they have it, they very often do not exercise it.

The freedom to choose a computer system supplier without becoming 'locked-in' is increasingly a requirement for many purchasers. The only way to obtain this is by forcing those same suppliers to conform to international standards so that families of systems are completely compatible.

But having obtained such freedom of choice, there are many good reasons why a customer may choose not to exercise it. These can vary from the favourable discounts that can accrue from volume purchasing to the potential higher level of technical support that may be obtained from a single source supplier. The competitive advantage for suppliers in the open-systems world will therefore change from technological differentiation to differentiation on the services they may offer.

Having looked across the information technology industry for the arguments for standards, we continue this chapter with a discussion of some of the counter-arguments.

WHY NOT STANDARDS?

Several arguments are used by antagonists of the open-systems movement to dissuade users from a move to standards in their procurement strategies. Common among these are:

- *Performance* 'Standard systems are slow.'
- *Cost* 'They are more expensive.'
- *Choice* 'There are too many choices.'
- *Progress* 'The definitions are incomplete.'
- *Migration* 'It is difficult to make the move.'
- *Security* 'Open systems are open to intruders.'

Any one of these criticisms could be made in certain circumstances against any computer system, proprietary or open. None is specific to, or characteristic of, open systems. In any case, each must always be viewed within the context of specific user requirements (both tactical and strategic) and against the external environment in which the user organizations must exist.

Performance

If a search is made for a single, optimum solution simultaneously to a number of different problems, the chances are that the solution reached is unlikely to be the best possible for *any specific one* of the problems. So it is with standards in IT.

A general standard developed to be usable across the almost infinite variety of problems for which computers are used today is unlikely to produce the best possible solution for a particular application looked at in isolation. Manufacturers with proprietary products optimized for specific purposes sometimes use this as an argument against open systems.

In highly specialized applications, maximum performance usually comes from modifying proprietary products, fine-tuning them for a specific purpose. If this performance is then compared with that for unmodified standards-based products, it is not surprising that the finely-tuned system comes out ahead. But in normal commercial applications, standards-based systems perform as well as, and often better than, their proprietary counterparts.

Users need to realize that the benefits of open systems lie principally in the long-term and strategic economic advantages and not necessarily in achieving the highest possible performance for an isolated application at the lowest possible price. If an application is particularly performance-dependent, it is quite possible that a customized solution, based on either standard *or* proprietary technology, may be appropriate.

Cost

The use of standard operating system technologies (such as UNIX for multi-user computers and MS-DOS for personal computers—both of which are discussed later) has driven the price of computers down dramatically over the last few years. Comparative pricing of such an open system against a proprietary one of similar performance will often show pricing down by a factor of two or more.

This is a logical effect of the shared technology which takes place in the development of open systems. Development costs are being carried by a *group* of companies rather than by one, as in the case of proprietary technology.

The potential adverse costs of an open-systems purchasing strategy come not from the costs of the technology itself but rather in the cost of *changing* from the old to the new. This '*cost of change*' is not specific to open systems; a change of procurement policy from one major manufacturer's proprietary technology to another's is likely to cost a great deal more than the cost of the change from proprietary to open. And once the latter has been undertaken, it will not have to be repeated in the future, even with a change of supplier.

In any case, these are short terms costs and must be measured against the eventual savings that will come from such things as: ready availability of know-how in the population at large; reduced learning times; reduced software maintenance and upgrade costs.

As the open-systems marketplace expands, the cost differential between open and proprietary systems will move even further in favour of open systems.

Choice

The development of standard open systems is in its infancy and there is more variation in the implementation of standards than proponents of open systems want to see. This variation has been reduced dramatically over the last few years and will continue, particularly as users start to understand the benefits of open systems and put pressure on their suppliers to conform closely to the standards.

The emergence of two standard operating systems (MS-DOS and UNIX), has dramatically reduced the number of variations in computer systems being sold and increased the number, quality and variety of software that runs on them. This trend will continue as conformance to standards becomes a requirement of user procurement policies and hence of their suppliers' products.

Progress

Standards are, and will always be, incomplete; technology is constantly evolving. Agreement on a standard usually starts at a base level, with pieces added in a quantum fashion as agreement is reached on each. This is an inevitable part of the standards-making process, which requires consensus from a variety of bodies around the world. This process will be discussed in detail later (see Chapter 4).

Sufficient push from the user community could force earlier commitment to standards from the computer manufacturers. In that case, the process of agreeing the additions to established standards could undoubtedly be speeded up.

Agreement on international standards for many components of the IT industry already exist and are supported by all elements of the marketplace. Accurate information needs to be made readily available to users so that their IT strategies are in line with mainstream standards activities.

Migration

It is difficult, and therefore costly, to change major installations from the use of one technology to the use of another. Not only do the new systems themselves have to be purchased and installed, but a substantive training and development program is usually needed.

Much of the software developed for the old systems has to be converted, or rewritten, for the new. And while all this is being done, the business, which often depends on its IT systems for efficient operation, needs to carry on uninterrupted.

The problems here are not specific to the change from a proprietary technology to an open-systems strategy; they apply even more for a change from one proprietary technology to another.

For easy migration, users need migration tools—software products that speed up the conversion process—and integration devices—products that allow them to link the old systems with the new and communicate between them. For the open-systems environment, there are many 'tools' to help with migration and integration of proprietary systems with open systems. These often allow both types to co-exist efficiently on 'mixed' networks of machines.

Clearly, it makes better sense to develop such migration tools for the proprietary-to-open path, rather than the proprietary-to-proprietary path, for which the volume requirements are much lower. Developers have recognized this by developing migration products for most of the previously popular proprietary products.

If retaining the status quo means that an organization is boxed into a corner a few years from now, unable to buy products or services except at prohibitive prices and cut off from easy computer communications with the rest of the world, then the costs of migration may turn into the costs of survival.

As an example, a large US organization which developed a major application system on proprietary computers suffered substantial losses when the computer manufacturer announced that the system used was to be discontinued. If the application had been developed on standard systems, the user could simply have switched suppliers, at minimal cost. Once the migration to open systems has been achieved, it does not have to be undertaken again, even if the user changes system supplier.

Security

All modern multi-user computer systems are expected to be secure to some extent. This usually means that they must provide some protection against unauthorized access to the system and a level of controlled access to sensitive data once initial access has been gained.

Open systems have been accused of a lack of security features following some widespread virus attacks that have taken place over large international networks. But this must be put into perspective.

First, the *only* large international networks that exist in a non-proprietary sense—i.e. to which many organizations and many different systems are connected—are based on open, standard technologies; without these, it would be very difficult to build and operate them.

Second, the ability to abuse protective mechanisms is not unique to open systems. All computer systems that can be physically accessed can be violated by sufficiently clever or persistent attackers. But since 'open systems' often imply very large networks with a mixed population of machines, any one of which may potentially be accessed from any other point on the network, the problems of security and control can become difficult to handle.

This is perhaps the only example where 'making IT easier' could be considered detrimental to the system as a whole. Making it easier to access, and therefore potentially to abuse, the systems make it harder to protect sensitive information. However, given that a system, whether proprietary or open, is designed with security in mind (and not, as is often the case, as an afterthought) this need not be so.

A great deal of work is under way worldwide to improve the management and control of secure systems. Much progress has already been made, spurred on by the needs of military and financial establishments. Today, security features available with open systems are at least as good as those in any proprietary environment.

Information technology for competitive advantage

We conclude this chapter with a brief comment on one of the most important issues facing organizations today—the proper management and use of information.

Often, the largest asset a company has is the information that is contained within it—information on its people, products, production schedules, inventory, customers, competition. Properly used, such information can lead to more efficient running of the organization within and much better targeting of its activities outside.

But the information in many organizations is highly distributed, often residing in the heads, or personal computers, of many people working in many different departments. The proliferation of personal computers and the networks of computers that have followed, has led to a mushrooming of departmental computer systems, often separate from the central, usually mainframe, large corporate system.

Each of these separate computers contains information on some aspect of the corporation. Often, much of it is duplicated. Information which might be useful to one department is not known or is inaccessible to it, being contained within the computer of another department.

Within the most enlightened companies, management is beginning to realize that if this information could be pulled together into one cohesive whole, and used intelligently, it could further the corporate aims dramatically. For this reason, the procurement policy for IT systems is now an intrinsic part of corporate business strategy.

Integrating the corporate information technology strategy with the business strategy is no easy matter. Not only is it necessary at a technical level to integrate the many diverse systems that already exist in the organization but it is also necessary to integrate the thinking of the IT managers with those that conventionally run the company. This is a communication problem of a different type and one that must be solved if the information management system is to be of maximum use in the pursuit of competitive advantage.

During the 1980s, many millions of personal computers spread within corporations worldwide. Individuals and groups developed applications software and used it to analyse information relative to their own business areas. Individual productivity was often considerably enhanced.

The next step requires that the machines holding this valuable information be linked together and the information contained structured into one cohesive whole. The technical problems of achieving this with many variations existing in the individual systems are huge. Obviously, the fewer variations that exist, the easier the problem of integration will be.

Fortunately, there is a universal standard already in existence for most personal computers, in the form of the IBM PC system and its clones. All other manufacturers, with the exception of Apple, supply products that are compatible with it. At the mainframe level, the standard is also set by IBM, the company which dominates the market.

At the mid-machine level, where the departmental, group computers belong, there is an emerging standard, based on the use of a standard operating system called UNIX (to be discussed in later chapters), supplied by AT&T. Compatible machines running UNIX are supplied by all the major computer manufacturers.

Thus with only three standard system designs—the IBM PC, UNIX systems, and IBM mainframe technology—it would be possible to construct corporate networks of the most general sort, incorporating many of today's already-existing systems. The problem of the integrated information system then becomes one of integrating the applications themselves, within which the information is contained.

The point here is that, unless the amount of variation in the hardware systems used is reduced, it is very difficult to integrate the information. If attempted, unnecessarily large amounts of technical expertise will be spent in integrating the hardware systems—a resource that would be much better used to develop ways of exploiting the information to the company's own advantage.

We give some examples of companies that have already realized the importance of these points when we present user stories in Chapter 8. As will be seen, the companies featured have taken the decision to reduce the variation in hardware systems in use within their organizations in order to concentrate their efforts on using them to better effect.

Note The increasing importance of this whole subject may be measured by the editorial inches it is now receiving in major publications, by new books published and by courses being run. As an example, in 1989, the US publication *Business Week* launched 'The Business Week Newsletter for Information Executives', with the message that 'subscribers will be the first to know which organizations are using information technology as a strategic weapon' and plans to offer 'advice on how to mesh diverse PCs, mainframes and global communications systems into a cohesive system to improve the business'.

Summary: standards—who needs them?

In this chapter, we have looked at the sectors that make up the computer industry market, and have briefly analysed each of their general requirements and some of their present constraints. By so doing, it has been possible to see needs which are common to most of the groups in the marketplace.

We found that every organization is seeking to make its people more productive, to reduce learning times, to hire people in from an already-trained reservoir, and to move them around in the organization with minimal need for training or retraining. This we defined as a need for 'people portability'.

Most are also trying to control and reduce product development times (and costs) and to provide better services to their customers. Some are trying to increase the target markets for their products without a proportionate increase in associated costs. They are seeking to improve their profitability or to reduce their reliance on single market sectors. We saw that many of the requirements could be met by a move to a few standard systems which allowed software to run unchanged on hardware systems from different manufacturers. This we called 'software portability'.

For the needs of many users, particularly those operating in departments or divisions, or trading with partners electronically, we saw a need for standard communications methods between computers, and concluded that these could more easily be implemented if there was less variation between the computers themselves.

Our overall conclusion is that, in order to accomplish better, more efficient use of IT and IT personnel, some of the *variation* that currently exists in the computer industry *must be reduced*. In simple terms, the computer industry suppliers must find ways to make information technology easier to use.

But what does this translate into, in terms of the technological changes required? That is the subject of the next chapter.

CHAPTER 3 The standards needed

It is clear from the last chapter that a widespread move to computer-industry standards, supported by all the major players in the industry, would enable users to address many of the issues raised so far. Indeed, this is the way the industry is going. But in the computer industry, standardization is a complicated issue and must cover an enormous number of different technical areas and functional requirements.

In today's environment, few departments, subsidiaries, or companies work independently or in isolation. For example, in computer-aided design projects the design team and the production group must work closely together. Draft specifications produced by one must be rapidly analysed for potential problems by the other. And when up and running, the production line must be integrated with information on orders and inventory, often in direct electronic communication with suppliers.

To handle such complicated, cooperative working situations, multiple systems must work in unison, able to communicate, and share data and tasks, with acceptable efficiency. However, as we have seen, there are many constraining factors currently limiting the growth and efficiency of IT usage in user organizations, many of which are due to fragmentation of the technology.

The only foreseeable way to improve matters is to take some of the unproductive variation out of the available products, so that, instead of spending vast amounts of time and resource on making incompatible systems behave in a compatible way, users can build the necessary distributed systems out of compatible components, and concentrate on using them to solve their business problems.

Although it is desirable to define standards throughout the computer industry, it is unrealistic to think that the requirements of the entire market can be satisfied with a single, simple standard computer. There is such a huge variety of application areas, user requirements, performance and price criteria to satisfy that, at best, one can only hope for a minimal set of standards to emerge over time. As technology evolves, it is to be expected that the standards will evolve too.

The objective must be to define standards for open systems in such a way as to allow innovation and competitiveness to continue among the suppliers of the technology, thus giving the widest possible choice of price and performance to the user. This is obviously an extremely difficult task, particularly when the stakes are high, and so many vested interests exist.

The overall standards required for the mixed networks of computers that are now common in many organizations can be grouped under the three general headings of:

- portability
- scalability
- interoperability

We will discuss each of these in turn.

Portability

In many companies or organizations, there is a clear requirement that the same application software package be able to be run on a variety of computer systems, even if these have been supplied by a number of different computer manufacturers. This should be achievable without any work on the part of the user, and preferably without requiring any modification in the software from machine to machine. All versions of the software should be identical and the output readily usable on other machines.

If this were to be achieved for a range of machines of differing price and performance, then purchasers would have the freedom of choice they require. At the simplest level this sort of compatibility and choice exists within the IBM PC-compatible range of machines.

A company could, for example, standardize on a particular word-processor throughout all its departments, on machines ranging from small to large, independent of the supplier of the systems. Clerical staff could move from department to department without needing to learn a new piece of software on each move. Having learned one particular package well, the knowledge could remain with the individual, even if the machine were later replaced.

If this could be done in such a way as to allow upgrades as new systems using the latest technologies appear, with the same application software packages supported, so much the better.

Furthermore, if the software authors were able to produce their software products on a variety of manufacturers' computers with very little effort, and were able to maintain a single version while doing this, their costs and those of their distributors and other suppliers would be reduced dramatically. In a competitive marketplace this also means better value for the user.

The requirement that the purchase and usage of application software be separated from computer hardware in this way means that application software must be easily portable across computers. It is this property that is known as *portability* of application software.

Scalability

Many organizations are made up of subsidiaries, field offices, corporate headquarters and departments of different sizes, with common applications requirements. In order to satisfy their IT needs in a cost-effective way, it is usually necessary to use computers of varying size and performance. However, if standardization takes place on the application software across departments, as is increasingly the case, it is desirable for the performance of the application to be acceptable on machines of different sizes.

The ability of the same application package to run with acceptable performance on computers of varying sizes, from the personal systems through to minicomputers, and possibly even to the corporate mainframe, is a second general requirement. This is the characteristic known as *scalability*.

Interoperability

Many organizations today have made sizeable investments in computer technology and in software written for the proprietary systems. As these organizations make the move to standard, open systems for the future, they will find themselves using complex, large and mixed populations of computers, often from very many vendors.

As the need for better use of corporate information and data becomes increasingly widespread, more and more of these computers will be interconnected. The corporate IT system eventually becomes a mixed vendor network, comprising, at least initially, both proprietary and newer, open systems. The number of proprietary systems can be expected to diminish over time as they are replaced, but not to disappear entirely. As the organization moves to standard systems in the future, the *investments* already made in IT must be *protected*, if at all possible.

To achieve this, communications methods between machines, both new and old, that provide accurate and reliable transmission of data without affecting the applications that are running in the separate computers, are needed. This is the requirement for interconnection.

Interconnection is only part of the story, however. As the computer system turns into a mixed vendor network, with many machines connected to it, it becomes feasible to move part, or all, of a job to other machines on the network when one machine becomes overloaded, or is not ideal for a particular piece of work, i.e. to *distribute* the software across the network. Eventually, users should become unaware of where in the network the application is running and this should be irrelevant to them.

In order to deal with the issues of distributed and shared software, it is necessary to combine the technologies for providing interconnection with the technologies that control the software applications that may be running on individual computers on the network. This is the requirement of *interoperability*.

As companies become flatter, more distributed structures, work is increasingly undertaken in groups, with decisions being taken by, and communicated electronically between, members of the group. New applications are emerging under the classification '*groupware*', which is software specifically designed for these shared tasks. Such applications will require that the systems used be interoperable.

Portability, scalability and interoperability

The two main issues of 'standard computer systems', defined through the concepts of (1) portability and scalability (PS), and (2) interconnection (I), are often treated separately. However, for many applications, they are no longer independent.

When systems were first connected over a network, users wanted them to communicate in order to move data from one to the other, or to enable them to share more expensive peripheral equipment, such as a laser printer or fax machine. Now, the emphasis is increasingly on full interoperability—the ability of machines not only to communicate but to share certain tasks between them. The evolution of this process is shown in Figure 3.1.

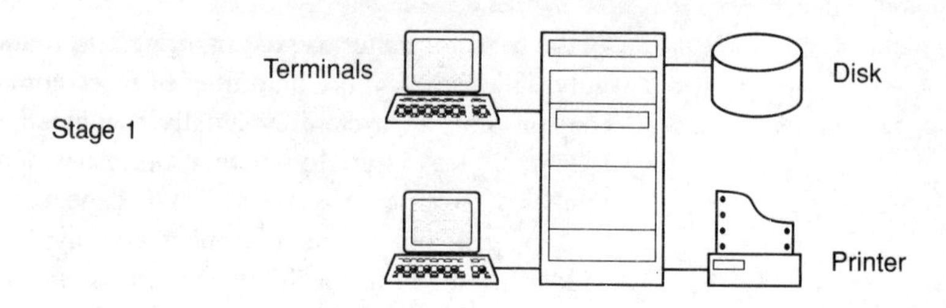

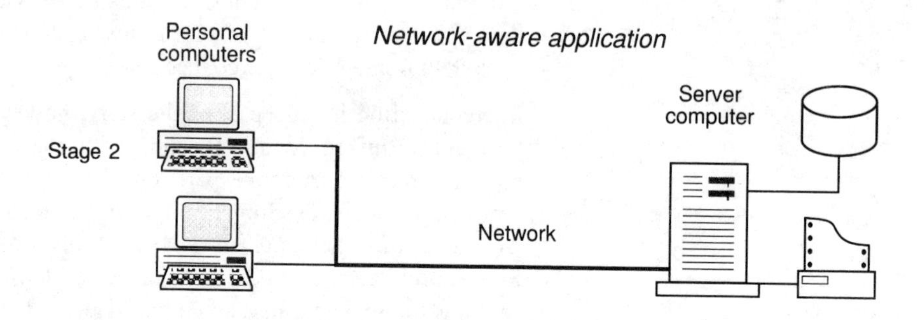

Network-intrinsic application
Personal computers
Computer server
Peripheral server
Stage 3
Network

Figure 3.1 Evolution of applications

At Stage 1, applications were originally designed to exist on a single computer, to which several terminals and printers might be attached. *Portability* and *scalability* are the important standardization issues in that context, where the desire of the users was for the same application to run efficiently on different machines.

Over time, at Stage 2, applications became 'network aware', meaning that although execution of an application takes place within a single computer, the application sees the network as a device for communication between machines and for sharing peripherals. The main concern is then for *interconnection* standards.

At Stage 3, applications become 'network intrinsic', in the sense that they can no longer be executed on a single computer—they need the presence and functionality of the other computers on the network in order to share the work. The technical issues in this case then centre on *interoperability*.

Such is true *distributed processing*, or *interworking*, where both machines and people work cooperatively in groups. Although few software packages exist yet for such 'groupware' applications, it is anticipated that many new and exciting software products will be built in the future.

To achieve interoperability, the application environment and the communication methodology must be considered *together*. As we shall see later (in Chapter 6), standards-setting activities for each, which to date have proceeded independently, are now converging.

The definition of open systems

Whereas in the US, 'open systems' is often taken to mean standards which address portability and scalability (usually through the use of the UNIX operating system), in Europe 'open systems' most often means interconnection (defined through the ISO/OSI standard, which partly addresses interoperability). Our definition throughout encompasses both.

When the three characteristics: portability, scalability and interoperability, are taken together, and international standards set for them by an open process, in which anyone may participate, and the results are available on equal terms to all, the result is to define that part of the computer industry known as 'open systems'.

Standards for portability

Levels of standardization

It is possible for standardization to take place at any one of several levels of a conceptual computer system (see Figure 3.2). Each of these has direct and indirect implications. We will discuss each in turn.

Standardization of the user interface (Level 7)

When using a number of application software products from several sources, it would be very useful if many of the attributes of the user interface—how the mouse works, what the graphics symbols on the screen mean, how the screen is laid out—were common. This would reduce learning times for new products and make the integration of one software product with another easier.

LEVEL 7	The user interface
LEVEL 6	Application
LEVEL 5	Software language
LEVEL 4	Software tool
LEVEL 3	Operating system
LEVEL 2	Computer hardware
LEVEL 1	Microprocessor

Figure 3.2 Levels of conceptual computer system

For this reason, standards attention has recently moved above the application level to the *user interface*—that part of the software that the user deals with directly. Following the success of the Apple Macintosh's user-friendly graphical interface, attention is now focused on defining a similar, but universally available, interface for open systems.

Although a standard interface is attractive for the users, producing one can create problems. Some of those companies which developed the original 'user-friendly' technology believe that the 'look and feel' of their interface designs should be protected under copyright law.

Clearly, a balance must be struck between the needs of the marketplace and the proper recognition and protection of original ideas and development investments. This could be done through appropriate licensing arrangements.

Developing an application for a specific graphical interface is a difficult task, and the pool of expertise available to do it is at present limited. Moving an application developed for one graphical interface to another is a heavy investment. Thus, the more of these interfaces there are, the more the costs will again proliferate in the software development and distribution chains. Standardization in this area is beginning to take place, with several major interfaces competing for acceptance.

Until a consensus is reached, software developers may postpone their development plans rather than risk losing time and money by guessing wrongly on which one will eventually dominate. This could be very frustrating to the users, who naturally want the best, most easily usable products available at the earliest possible time.

Standardization of the software application (Level 6)

Organizations have achieved considerable savings in software development costs and in the costs of training and retraining of personnel through a decision to standardize on certain applications used throughout the company.

For example, standardization on a word-processing package such as WordPerfect, which is available on machines of different sizes and from many manufacturers, allows a wide choice of hardware systems to be purchased, suitable for departments of various sizes. A secretary trained on WordPerfect can be moved from a PC to a terminal on a machine supporting, say, 50 users, and yet be able to operate exactly the same word-processing software.

Standardization at this level gives the purchaser freedom of choice of hardware supplier and at the same time allows staff to ignore any special features of the computer environment which may lie beneath the application software.

Superficially, standardization at this level provides many of the advantages that the users seek with 'open systems'. However, as we have seen earlier, unless standardization takes place at a deeper level, there are many invisible costs both to the software authors (porting work, administration of different versions of the products) and to the distribution channels (costs of inventory, technical support), that will undoubtedly have been passed on to the user in the price of the product.

Furthermore, because the product must be maintained in many different and proprietary system environments, the technical support costs can be unnecessarily high when things go wrong, or when product updates occur.

Standardization of software languages (Level 5)

Since the development of the computer, standardization work has proceeded for software languages in common use. Thus there are American National Standards Institute (ANSI) standards for widely used languages such as COBOL, BASIC, FORTRAN, ADA and, more recently, the C language.

Standard languages allow application developers to write to a common definition, knowing that their application should run on any machine that supports *that* version of *that* language.

Unfortunately, few suppliers of these 'standard' languages have been able to resist adding 'enhancements' to them, in the belief that this 'adds value' to their implementation.

While this could be true if the enhancements suit a particular application, it is not a policy that adds value for the industry overall. Once again, such differences create costs throughout the distribution network, which are ultimately reflected in the prices that users pay. If such special features *are* generally useful, then they should be incorporated into the standard definition of the language.

Languages such as COBOL or FORTRAN are often produced by independent software developers who supply them to many hardware manufacturers for onward sale. Because the porting cycle often takes months, if not years, to cover the major machines, the most recent version of the software cannot generally be made available to all manufacturers simultaneously.

When supplied with the software, the hardware manufacturers usually test the language on their own computer systems, debugging the software and customizing the documentation as they go. Some manufacturers achieve this in weeks, others in months.

The end result of this process is that, even if the same software *were* supplied to the hardware manufacturers at the same time, the products available to the users in the field at any one time from different suppliers, are likely to be different.

A major supplier of applications software for accounting went out of business a few years ago when the users of the software experienced a wide variety of problems that were initially attributed to the application packages. A post-mortem revealed that the problems were due to the largely unpublished and invisible differences in the versions of the 'standard' language in which the applications were built. These differences caused the accounting package to fail unexpectedly, which in turn caused the supplier's business to fail.

Standard hardware systems will not completely solve problems such as these but they would go a long way towards doing so. The porting cycles for the developers would be greatly reduced, perhaps allowing them to spend time and resources more constructively, improving the overall quality of the software products. Hardware manufacturers might then not be forced to 'improve' the products and so introduce many of the differences experienced today.

Standardization of the software tools (Level 4)

As the cost of development rises, organizations developing customized software solutions have moved to the use of 'software tools'. These are designed to reduce programming development time and costs. Chief among them are database management products, application generators and fourth-generation languages (4GLs).

Increasingly, because of the fundamental importance of the corporate data, more and more organizations are choosing to standardize on the database management system used to control that data within their company. Since it is usually necessary to produce reports from analyses of the data, standardization often takes place also on the application generator or fourth-generation software product used to produce these.

Although this approach does allow for portability of people and know-how around the organization, and for a degree of interoperability between the databases held on different machines on the network, it again suffers from the fact that many hidden costs are still inherent, albeit pushed back to the software authors, distributors and suppliers.

A common criticism by the users of standardization at the database and 4GL level is that it has simply moved the 'proprietary lock-in' higher up the computer system. In other words, the users are still locked in to technologies that are specific to one manufacturer—in this case the database supplier—and are prey to the vagaries of pricing and product development strategies of that single supplier.

To overcome this, at least in part, standard methods for access to data contained in databases are appearing from a variety of vendors. Based on technology originating from IBM, the access method known as SQL is now built into the best-selling databases for multi-user machines. To date, little or no standardization exists among the popular 4GLs, although users are increasingly seeking this.

Standardization of the operating system (Level 3)

The operating system software is that software which manages the higher levels of software and drives the computer hardware itself. It is the lowest level software and is intimately tied to the hardware on which it runs.

At the same time, higher levels of software, such as languages, databases and applications are intimately tied to it. Thus, applications software which runs on one computer will not generally run on an identical hardware system running a different operating system, without major modification to the application.

One way to achieve portability of software across hardware systems is then to standardize on the operating system itself. This is the approach taken by licensees of the UNIX operating system, which is used today by many manufacturers of machines, from personal computers to the largest mainframes, worldwide.

Unfortunately, although the UNIX operating system could, and some day undoubtedly will, provide the portability that the world requires, to date it has not managed to do so, for much the same reasons as have compromised the concept of 'standard' languages.

Manufacturers of computer systems have added their own features to the operating system software as supplied to them by the developer, AT&T, and so have destroyed the very portability requirement which the market needs.

It is ironic that these differences are often described by the computer manufacturers as their 'added value', for each of them *subtracts value* from the prices the users pay for software and support by causing them to incur unnecessarily high prices for software development, packaged products, training and technical support.

There are some instances in which it has been necessary for manufacturers to add features to UNIX for particular markets—security for defence applications, transaction processing for particular commercial situations and so on. Some manufacturers have chosen to specialize in such markets and developed the operating system accordingly.

This is quite justifiable modification provided that ultimately such features are incorporated into the standard operating system on offer to all. This is currently in process through cooperative efforts by a number of manufacturers in many important areas (see Chapters 4 and 7).

Standardization of computer hardware (Level 2)

The simplest possible approach to computer standardization is epitomized by the IBM PC and compatibles market. Here the strategy of the other computer manufacturers has been quite simple: 'Given that there is a machine selling in high volume, we will copy it completely, top to bottom, hardware and operating system software.'

This approach, while not technologically innovative, has produced (1989) an installed base around the world of more than 30 million completely compatible computers, in the sense that a disk containing software for one system can be inserted into another and guaranteed to run, independent of manufacturer. This is exactly the situation that exists in the music industry, where compatibility exists across hi-fi systems for compact discs or cassettes.

This enormous base of installed personal computers has in turn produced a huge population of people who can move around and across it. It has inspired a talented software industry to produce a wealth of the most creative software. It has driven prices of these computers and associated software products to unprecedented lows. It has certainly demonstrated the advantages of compatible systems to the users and to the software developers.

But what has it meant to the hardware manufacturers? Many of them have found it difficult to ride the PC wave. Used to high margins and until now not overly concerned with maximizing efficiencies in production, many have found the PC revolution a mixed blessing. Some, like Compaq and Amstrad, have taken tremendous advantage of it, either by improving on the technology without destroying compatibility (Compaq) or by efficient manufacturing and distribution, so reducing the price (Amstrad).

A common criticism of this top-to-bottom 'cloning' approach is that it stifles creativity and innovation. Even at its launch, no one claimed that the IBM PC was leading-edge technology. And although the software industry benefited initially by the ability to target to a standard base, it has become increasingly frustrated by the technical limitations provided by the PC hardware and software platform.

Technologies exist today that would enable software to be built that would be much easier to use and more powerful by far than current PC products. Unfortunately, today's PC, with its rather primitive operating system software (MS-DOS), cannot run them. Change is therefore inevitable, but it must be designed so as to retain the advantages of the PC approach, including as much as possible of the users' investments in software and expertise. For this reason, it is very important that the path to improved technology is such as to allow the current software products to run.

Standardization on the microprocessor (Level 1)

It is not possible for application software written for a computer using one microprocessor to run completely unchanged on a computer using another. The source code (the raw language in which the product is written) can remain the same, however, if the operating system software is standardized. This is *source code portability*.

When the microprocessor is changed, the source code must be translated into the set of instructions that the new microprocessor uses. This process is 'compilation'.

One can only have portability *without* recompilation between machines that use the *same* microprocessor *and* that have the *same* operating system. Such portability is referred to as *binary portability*.

The IBM PC and compatibles are binary compatible, being completely identical in classes based on the same Intel microprocessors. However, PCs based on earlier Intel chips, such as the Intel 80286, are *not* binary compatible with those based on the newer Intel 80386, nor will they be with those using the Intel 80486.

Software which uses the more advanced features of the newer processors will not run on their predecessors. Software written for the older systems, on the other hand, will generally run on the new. Consequently, software authors cannot have portability to all PCs; they are constrained by old technology. If they write software for the old systems, it will run on the new, but without using any of the new features. If they write for the new systems, the software will not run on the old, large installed base, of systems in the population.

Moving to machines other than the PC clones, the ideal would allow a choice of microprocessors of differing characteristics and from different manufacturers to be usable within the standards otherwise set. Microprocessors could then be designed and built to specific performance requirements while still allowing conformance to source code portability standards.

There is a move in the computer industry to define *Application Binary Interfaces (ABI)* standards, to give binary portability between machines using the same microprocessor, and source code portability otherwise. Manufacturers of systems who follow these design standards will be assured of producing compatible machines. ABIs are currently being defined for microprocessors from Intel, Motorola, SUN Microsystems, MIPS and others.

Standards for interconnection

Today, data is distributed throughout organizations on systems of varying functionality, from personal computers through departmental systems to the corporate mainframe. Many companies face the problem of how to bring these together into one cohesive information system that can be managed efficiently for the purpose of improving the business. This mean devising methods for communication between different devices within and between various departments and, as organizations become increasingly international in scope, possibly across international frontiers too.

To reduce the complexity of the design and management of emerging network configurations, the computer industry must agree international standards for communications between devices. Manufacturers can then incorporate such standards into their products and purchasers may specify them in their procurements.

Until recently, the approach taken by many users was to buy as many of the needed products as possible from a single vendor, who then had to provide the necessary communications for all devices supplied. If particular equipment was required that the vendor could not provide, the user had either to manage without or do some customized (and expensive) development work in order to attach the new devices.

Large vendors, such as IBM with its Systems Network Architecture (SNA), have well-established communications technologies. Other manufacturers can use SNA technology to attach their own equipment to an SNA network. Few other manufacturers have adopted SNA within their own networking products and so SNA has not become an industry standard.

Today, these manufacturers and their customers often have huge investments in proprietary technologies, making if difficult for them to move to open standards easily. Because of this, users are building open networks side by side with their proprietary ones and constructing gateways between them. (A gateway is a device which connects two dissimilar networks and performs the physical, electrical and protocol translation between networks.) This is a necessary cost today but one that will tend to decrease in future as the world moves to networking standards.

In the same way that departmental systems within organizations no longer stand in isolation but are connected into the corporate network, so also are organizations becoming connected to each other for specific purposes. With the growth in electronic trading, companies are requiring their customers and suppliers to conform to specific formats in the communication of information. This area of the technology is electronic data interchange (EDI). In addition, more and more companies are linking up their systems to public or private information systems, some of which are now worldwide in scope.

The fewer the number of different systems that exist on the network, and the more universal the communications standards, the easier will become the distributed processing, systems administration and network management issues. As standards become universally used, the greater will be the knowledge of standard technologies in the population at large. This is going to be a vital issue in the future, as these systems grow larger and more complex.

To illustrate the size and international nature of the potential networks of the future, consider the example provided by Usenet, a large information system which has evolved over the last few years.

Usenet: an example of a large multi-vendor network

An example of a very large international communications network, Usenet, has evolved over the last few years. Often referred to by its users simply as 'The Net', it provides a free-form electronic forum for exchange of information worldwide. The Usenet facilities are run on a non-profit basis and are usually free to users, apart from the costs for the modem, telecommunications software and line charges incurred.

Usenet currently connects more than 7000 computers in more than 30 countries and provides more than 250 000 users with up-to-date information on technical, marketing and miscellaneous matters. No one knows precisely how big Usenet is, nor how many people use it. Anyone with a UNIX system and a modem is free to join it, provided a nearby site has set up as a major node and agrees to the connection going through it.

Usenet is a source of technical information on the computer industry, primarily, but not exclusively concentrating on that part which is developing open standards based on the UNIX operating system. Users can ask questions over the network, and obtain help from other users, wherever they may be. Usenet is not an electronic mail system; anyone may read any item on it.

To date, Usenet has no overall network management, or manager. It has evolved over time to its present size, with users so far behaving more or less rationally and reasonably. However, it has begun to show signs of strain, as it grows ever larger, and it is obviously prone to attack from 'viruses' and other forms of abuse, in part because of the ease of access.

Usenet has now assumed such importance in the computer industry that steps are being taken by interested parties to ensure its survival and healthy growth for the years to come.

Open systems: further requirements

As we shall see in Chapters 5 and 6, much of the work needed for development in developing standards for open computer systems is well advanced and achieving widespread acceptance. But there are specific areas in need of further detailed work. The following sections describe those of immediate importance, in overview. More detailed descriptions and discussion of these are contained in the Appendices.

The need for bridges

In order to preserve as much as possible of prior investments made in proprietary systems, a move to an open systems procurement policy leads, in the first instance, to a requirement for 'bridges' between the proprietary (present) systems and the open (future) standard systems.

These 'bridges' may be in the form of migration tools that aid the process of moving software from proprietary to open systems—for example, a software tool that converts software written in a proprietary language (such as RPG) to a standard language (such as C).

The 'bridge' might be a specific communications device that converts software as understood by one machine to that understood by another—for example, documents prepared by one particular word-processing system could then be sent across a network through a translation device to an editor to edit on an otherwise incompatible system.

Other 'bridging' requirements exist in the form of customized training programs, designed to convert knowledge of a proprietary system into the equivalent knowledge in the open environment. An example could be a UNIX training course designed specifically for people who already have experience with Digital's VMS operating system.

The need for integration

In the move to open systems, users will need to integrate various pieces of hardware and software, both open and proprietary, and from a number of different suppliers, into a cohesive IT system, designed for their own particular needs.

Some organizations have already developed extensive system integration skills internally to handle this, but others will need external professional help. Many third party organizations, including divisions of some of the large hardware manufacturers and traditional consultancies, have recognised this and are gearing up to offer such services.

The need for integrated network management

Until the new open systems dominate the market, which will not be until the mid-1990s, users will be running increasingly complex, mixed-vendor and mixed-technology networks. These will require standard communications between the machines on the network and between the software applications running on them.

Today, not only are there many different pieces of equipment on these networks but similar equipment from different (or even the same) suppliers are incompatible. Furthermore, each manufacturer supplies their *own* proprietary tools for the administration and management of their *own* systems and networks. This can lead to situations where the (human) network manager must watch several terminals, one to monitor each of the proprietary systems on the network.

As these corporate networks become more complex and use of the information contained within them more vital to the company's business, the job of managing the network is becoming more and more difficult. As a 1988 report from UK IT management consultants, Butler-Cox, says: 'Most network managers today are forced to use a collection of discrete and incompatible network management tools, often dealing with as many as ten different suppliers.'

Ideally, network managers should have an *integrated network management system*. This would take information from all the different makes of equipment on the network and process it, giving the operator a complete picture of the network on a *single screen*. From this screen, the manager should be able to diagnose faults, measure performance and manage traffic flow along the network.

Although there are many network management tools on the market, none of them so far meets these criteria. Instead, they are usually targeted to specific manufacturers' proprietary products. Fully integrated management tools will be vital for open-systems networks.

The need to define the system administration job

Responsibilities for system administrators and network managers vary widely among computer installations. In some environments, the job is defined as 'keeping the computing services running efficiently for the users', which encompasses everything from hardware maintenance to network management, and could include user support and consultancy. In others, the task of systems administrator is more rigidly defined—for example, to control system access and administer software back-ups.

Before standards can be defined for systems administration tools, the specific areas to be controlled by the systems administrator must be defined and agreed. In other words, standardization of the job description is needed before standards can be set for the tools to implement the tasks.

The standard definition for network management so far describes five functional areas: configuration management, performance management, fault management, accounting management and security management. These are all being addressed by standards bodies at present.

Note There is likely to be a resurgence in external '*facilities management*', as companies come to the conclusion that they need expert help in order to manage these very complicated distributed computers. This could well be a role assumed by the systems integrators.

The need for security

Since the birth of the computer, security of access has always been an issue. If it is remembered that the early computers were often dedicated to cryptography and used in military establishments, this is not surprising.

In the early days, secrecy was the vital requirement. Today, security is much more concerned with restricting access to certain data only to those users authorized to see it; verifying that users of a system are in fact, authorized; preventing sensitive data from being accidentally disclosed when on route between systems; preserving the integrity of messages sent; and proving that messages have in fact been received.

Related areas clearly involve issues of encryption of information and methods for digitizing and verifying identifiers such as personal signatures.

Open systems are sometimes criticized for being unduly open to security abuse. This has little to do with the adoption of open PSI standards. Rather, it is a result of the fact that, as more and more systems are linked together, it becomes *physically* possible to access many systems over the network that were previously isolated in secure geographic situations.

A valid criticism could be that with the use of common standards, it is easier for a potential intruder to become expert in the technology. There will ultimately be more people with the necessary skills. This is perhaps the only example where a build-up of computer-related skills in the population at large may not be altogether a good thing.

Standard algorithms for the encryption of data have been in use for some years. The US Federal Information Processing Standard (FIPS) for security was first published in 1977. Given the performance of computers today, these algorithms are no longer sufficiently complex to prevent the codes being broken by the serious attacker and much research has been done to improve them. In general, these no longer rely on mathematical subtlety. Instead, methods have been developed which allow decryption, but only over such a long period as to be of no practical use. Clearly, such methods then *have to change continually* as technology advances.

Access control standards also exist, again driven by the US government through the activities of the National Institute of Standards and Technology (NIST) in the US, the body responsible for the FIPS standards for federal procurements. Contained in *Requirements for Trusted Systems*, known by its cover as 'The Orange Book', various levels of security procedures for access control are defined. These standards must be incorporated into products as part of the open (PSI) standards.

The need for conformance testing and branding

For most procurements policies, it is not enough for manufacturers to state that they provide compliance with required standards. Purchasers also require *independent testing of the claimed conformance*.

As standards are set in various areas, products for testing manufacturers' implementations of them must therefore also be developed. Easy identification (branding) of those products that have passed such tests is useful and so standards for *branding* must also be defined.

The need for international standards

Throughout the 1980s, the market for computer products in the US was buoyant, and the major manufacturers concentrated on it. As we enter the 1990s, the US market appears to be slowing down, while the international market appears to be accelerating. This is placing an increasing emphasis on international issues. Added impetus comes from the large number of organizations, both commercial and other, that are increasingly seeking to operate multinationally. In all, there is a need for 'internationalization' of open systems.

Internationalization can be defined as the provision of tools and utilities to facilitate the use of IT products in countries using different languages. For example, the provision of facilities in the computer operating system, to handle characters used in languages other than English, is part of the internationalization process.

Internationalization should be distinguished from *localization*, which is the requirement to provide language-specific and local custom enhancements. For example, the ability for a particular accounting software package to handle different accounting practices for various countries is a localization issue.

The objective of internationalization is to enable users to operate systems and applications in the local language without having to do any extra work.

Since most of the technical development of products for the open systems marketplace has so far been in the US, this problem usually translates into a need to operate systems in languages other than English. This means internationalization not only for the products themselves (hardware and software) but also the materials (documentation, promotional marketing materials, demonstration products, training manuals) that surround the products.

Suppliers eager to expand into international markets have individually tackled the problem of internationalization in a variety of ways, with the consequence that there are many variants in the marketplace. As with other proprietary enhancements, these now need to be brought together to provide one international standard for open systems.

International issues

Today, the centre of innovative research for both hardware and software is in the United States. This is illustrated by the fact that the cross-industry software—spreadsheets, word-processors, databases, 4GLs, languages, communications products—that dominate the international software market, originate in the US. These are widely used by systems and software houses in Europe, and other parts of the world, to build localized, vertical applications for specific users and markets. There are many reasons, both economic and sociological, for this domination by US suppliers.

The time for, and costs of, development of software are approximately the same anywhere in the world (with the exception of the software development 'factories' emerging in, for example, India). The costs of producing support materials, such as manuals, marketing brochures, demonstration software, are likewise comparable. The costs of producing and running an advertisement in a major computing magazine do not vary greatly across most countries, although in the US they can be higher than elsewhere.

But the size and uniformity of the market reached by any promotional activity in the US is many times that of other national markets. This gives US software companies the advantage not only of early access to new and innovative hardware technologies developed locally to them, but also a large, homogeneous home market in which to reach critical mass before expanding abroad. This is not the case for developers in any other market.

A UK, or other international, software company *must* expand outside its home market if it is to reach any credible size, but to do so, it faces vast obstacles. If it expands to the US, the direct costs of set-up and market entry are large, and the risk of failure in a foreign market is high. If it expands into international markets other than its own, it is immediately faced with enormous costs of translation and localization of the software, or other, products, before having reached sufficient strength in its home market to support the overseas investment.

This situation applies for any non-US software company. The result is that there are few high-volume, Europe-originated software products selling in Europe; the same applies in other markets. Even fewer products have crossed the Atlantic (or Pacific) back into the US with any degree of success.

As a consequence, there are almost no software products available for the open-systems environment in languages other than (American) English. There is also very little expertise available anywhere in the world in the technical issues relating to designing software for the international market.

Although there is a vast amount of activity in the various technical committees around the world that are working on international standards, not much practical information seems to have filtered out to the software industry as a whole. Although there is venture capital investment money available for development of technologically innovative products, no development capital appears to be available for solving the problems of producing products for a truly international software industry.

Superficially, this position might seem advantageous to the US suppliers. However, this is a narrow view. Unless there are products which satisfy international needs, growth of the international market will ultimately be constrained.

Europe: open frontiers?

In 1992, Europe is to become a barrier-free market, allowing people, goods, services and capital to move freely across country boundaries.

The European Community (EC) currently comprises twelve countries, representing the interests of more than 320 million consumers. This is a larger market than that of Japan or the United States. Reflecting the potential in the European market, US and Japanese companies are increasingly focusing on it. Local companies at the same time are expanding across Europe, usually by acquisition.

With the approach of 1992, 'open' in Europe will expand to include *open frontiers*, while the requirement for portability of people and applications will assume the additional dimension of portability across international frontiers. This raises some interesting issues for the open systems segment of the IT industry.

The fact that so much documentation on information technology products is *only* available in English, combined often with the availability of software in English only, is having the effect of slowing down the acceptance of some products and technologies. On the other hand, it is interesting to note that this is also having the effect of speeding up the acceptance of *English as the standard language, worldwide.*

For example, since the X/Open (see Chapter 7) literature is currently available only in English, an interested party has either to ignore this important standardization work or become fluent in English. X/Open plans to produce translations of some of its documentation, starting in 1990.

Japan Inc.

While internationalization of products is required for the European market, there is commonality in many of the character sets used. In Japan and other Far Eastern countries, this is not the case. The problems of handling the Japanese character set are complicated and those of dealing with traditional Chinese, as used in Taiwan, even more so.

Nevertheless, many Western manufacturers wish to sell into countries in the Pacific rim and, of course, the Far Eastern suppliers want to sell into Western markets. Neither will be able to do this without standard methodologies to handle the character sets. These must be agreed by the manufacturers, and built into standard products, with compatibility maintained between manufacturers.

In order to sell in high volumes abroad, Japanese and other manufacturers must increasingly conform to all internationally agreed standards. With the emergence of such standards, companies in the Far East may also more easily address the international market for applications software—an area in which they have so far had limited success.

Summary: the standards needed

We have seen that many of the factors currently limiting the growth and efficiency of IT usage in user organizations are due to fragmentation of the technology and lack of attention to true worldwide market requirements.

The needs of many users of IT could begin to be met if the computer industry were to adopt industry standard, open systems, based on *portability, scalability and interoperability* (PSI) on which applications for the international market could be built.

When the three characteristics of PSI are taken together, and international standards set for them by an open process, products supporting PSI could be built by any manufacturer who so wishes. These products must be able to be tested for conformance to the standards by independent bodies, and branded as compliant.

The predictable consequence of a decision to move to an open-systems procurement policy is a requirement for early and specialized information, technical support and training. To preserve as much as possible the investments made in current proprietary hardware and software systems, 'bridging' products and sophisticated integration skills will be needed. As networks become more complex, integrated network management and systems administration tools will be required.

For the global market, international standards need to be agreed. Open systems will open up frontiers but will also open up markets to new competition.

But . . . These standards *will* come and they *will* ultimately dominate the marketplace. There is now too much momentum behind them, and too many users who want and need them.

Any user organization that does not take steps *now* to design an open-systems strategy, and start to implement it soon, is likely to find itself boxed into a corner in the future—unable to hire people with the expertise to maintain its proprietary systems, unable to buy software products to run on them and unable to communicate easily with the rest of the world.

CHAPTER 4 The making of standards

In previous chapters, we have looked into the general requirements for standards in the computer industry. First, in Chapter 2 we considered *who* needs such standards and *why*. That was followed in Chapter 3 by a discussion of *what* particular aspects of IT configurations need to be standardized in order to start to achieve the desired objective of 'making IT easier'.

Before proceeding further into the technological detail of the required standards and examining those that currently exist, it is instructive to consider *how, and by whom*, standards are set. Without a basic understanding of this, it is difficult to put much of the present activity in the standards arena into its proper prospective.

This chapter first discusses the various groups associated with the setting of standards, following this with a description of the two main types of standards, public and *de facto*. It finishes by taking the reader through the detailed structure of one such standards body, the US IEEE.

Groups that set standards

There are many organizations in the world that have an important influence on standards evolution. They can be characterized as public, user or industry groups. The *public groups* are formal organizations, usually funded by governments, to define precise specifications for standards. The *user groups* can be made up of users from particular bodies, such as government agencies, with a well-defined role in the process, such as procurement specifications, or they may be groups of individuals, organized to take a particular part in the standardization process. The *industry groups* are vendor organizations working in areas of mutual interest and sometimes cooperating on development work, perhaps with input from the users. All three of these groups have a part to play in the process, and they cooperate and interact, one with another.

We will look at each of them, relating their activities to standards setting processes, and to the products under development for implementation of standards.

Public standards groups

Examples of public standard groups can be found in almost every country in the world. Prominent among these for the setting of IT standards are the International Standards Organization (ISO), the American National Standards Institute (ANSI), the US Institute of Electrical and Electronic Engineers (IEEE), the European Committee for Standardization and the European Committee for Electrotechnical Standardization (CEN and CENELEC), and the Japanese Institute for Standards (JIS).

Membership of public standards bodies is usually open to anyone that can afford the time and costs of participation, both of which can be considerable. In the IT industry, the work therefore often relies quite heavily on voluntary labour, sometimes donated by parties having a vested interest in the outcome of the work. This means that the procedures for defining the standards have to be set in such a way as to make sure that no special commercial advantage is obtained by those companies that donate the resources.

Industry groups

A feature of standards activity in recent years has been the emergence of various industry bodies formed not usually to write standards but rather to endorse existing ones and to help in the birth of more by defining concensus requirements.

Membership is usually fairly open but costs vary widely. Some have several classes of membership, based on size, or 'ability to pay'. There may also be criteria based on the ability to contribute productively to the required work. Sometimes, the body consist of members, all of whom are trying to solve the same, or similar, problems.

Industry bodies may be professional, well-funded and with full-time employees (or sub-contractors). Examples are the X/Open Company, formed by the major hardware manufacturers to define open systems (see Chapter 7); Unix International (UI), formed by AT&T and its UNIX licensees, to define and control the future of the UNIX operating system; the Open Software Foundation (OSF), a software company set up and funded by some of the major hardware manufacturers, notably IBM, Digital Equipment Company (DEC) and Hewlett Packard (HP), to produce products that meet international standards and are intended to become *de facto* standards in themselves; the Corporation for Open Systems (COS), set up and supported by representative organizations from all parts of the computer industry to define the standards needed for open communications between machines.

Bodies may also be lower budget activities, with minimal full-time or voluntary labour. Examples of these are: Uniforum, the commercially oriented UNIX user group, (previously known as /usr/group), responsible for much of the early work which led to emerging open-system standards; Usenix, the more technical, parallel organization to Uniforum; EUUG, the European UNIX User Group, which cooperates with both Uniforum and Usenix; JUS, the Japanese UNIX Society.

User bodies

Large users of information technology usually set data-processing standards for their own internal use. Usually they prefer to base these on public, officially approved, standards and they usually like well-defined standards, without any variable or optional features.

Government and military organizations dominate this group. Examples are: the US NIST (National Institute for Standards and Technology) which sets procurement guidelines for the US Federal Government and which has an equivalent in most other countries, and NATO, which sets military procurement guidelines in Europe. Major commercial organizations, particularly those whose businesses depend on IT, are increasingly active. Examples are General Motors and Boeing Aircraft, which have set communications standards for manufacturing and office automation applications; Eastman Kodak, which actively participates in many standards activities in order to make sure that its corporate standards are in the mainstream; DHL, which, because of its widespread international activities, needs to ensure that whatever it does is applicable to all the countries in which it operates; Visa International, which is concerned with standards for security and electronic data transfer together with most other multinational financial institutions.

De facto and *de jure* standards

There are two major types of standard: *de facto* and *de jure*. These Latin phrases mean, respectively, 'as a matter of fact' and 'according to law', which suggests that *de jure* standards have the force of law, while *de facto* standards do not. In fact, while there are an increasing number of de jure standards affecting information technology, no offence is committed if they are broken or ignored: they are not backed up by criminal or civil legislation.

The legal aspect of *de jure* standards exists because they are produced by a body with legal status, sanctioned by a national government—or federation of national governments—to produce standards. In order to create a standard, these bodies must use a procedure called *open process,* in which any person (or company, or country) with an interest must be allowed to comment on drafts of the standard. All comments must be responded to by the working group responsible for the standard. This is a public process, and consequently, *de jure* standards are generally known as *public standards* and set by public bodies.

Eventually—possibly after several years—consensus is achieved, and a democratic vote results in the standard being accepted. Due to the inevitable combination of interests involved in reaching consensus, *de jure* standards invariably take longer to come into use than their *de facto* counterparts.

De facto standards do not have to observe such diplomacy in their production. Indeed, many *de facto* standards are not the result of a conscious effort to produce a standard. Instead, they originate as innovative ideas or attractive products which just come to be accepted as standards in the course of time. This is the origin of the term *de facto*. It describes a standard which exists in fact, whether or not its originators had intended to create it.

Producing *de facto* standards

De facto standards emerge from products which are successful in the marketplace. These become standards through widespread use. There are several variants in the process.

Case 1: single product to single standard

In the simplest case, an innovative product is developed, sells well and becomes accepted in the marketplace before competitive products, which may do the same thing but in a different manner, can achieve significant sales. Eventually, when market penetration is high, a public body may review the technology and approve it as a standard (see Figure 4.1).

An example of this process is provided by the audio tape cassette, introduced by Philips in 1964. After a slow start, this was established as a *de facto* standard. Eventually, in 1983, the International Electrotechnical Commission (IEC) accepted it as a public standard and the compact cassette had officially arrived. In practice, it had arrived years before, but it took time for the official standards-setting process to catch up.

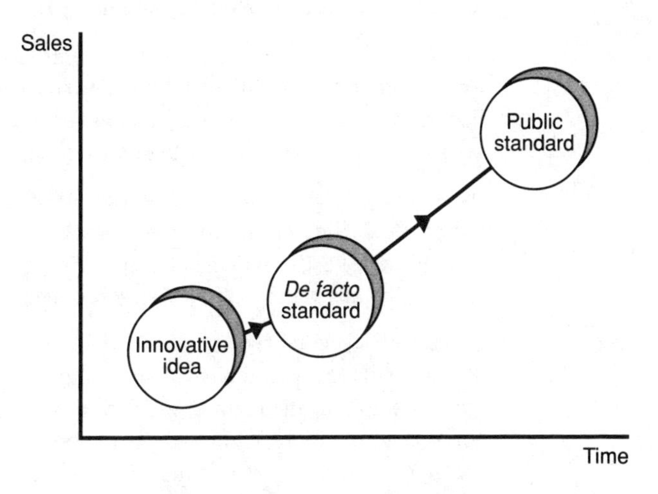

Figure 4.1 Single product to single standard

An important point arises here. Philips created the compact cassette, and protected its design through patents. It could have remained the sole producer but chose instead to license the technology at low cost to others. The hope was that a choice of suppliers would encourage consumers to buy more cassettes and, of course, the equipment that is needed to use them. Even now that cassettes are defined by an international standard, Philips still has the right to collect royalties from other manufacturers. The existence of a standard implies that the technology may be used freely, but not necessarily for free.

It is possible to pursue a strategy of this sort without realizing it. For example, in 1983 Hayes Microcomputer Products introduced a modem which accepted mnemonic commands beginning with the letters AT to control auto-dialling and other functions. It sold well, and developers created software packages which took advantage of its flexibility. The command structure was soon imitated by competitors wanting to take advantage of the new market opened up by Hayes. Hayes, too busy meeting the demand for its product, did not take steps to protect the commands and hence the ability to collect royalties for their use.

As a result the commands became a widely used *de facto* standard, and Hayes has so far failed in its attempts to exact payment from other companies for their use. There is, as yet, no public standard covering the Hayes commands.

Single or multiple suppliers?

The word 'standard' is sometimes taken to imply 'open' in the sense that a product which meets the standard should be available from multiple sources. This must *always* be true for public standards, because public standards bodies insist that there is no restriction on the use of any part of a standard.

Many public standards embody the intellectual property of individuals or companies in the form of patents. This is acceptable to a public standards body, provided that the owner of the patent gives an assurance that it will be licensed at a reasonable rate to anybody who needs it.

For *de facto* standards the position is quite different. The owner of the intellectual property on which the standard is based is under no obligation to license it to anyone. The result is that some products which have come to be seen as *de facto* standards are available only from one supplier.

An example of this is provided by the PostScript page description language for printers and visual displays, developed by Adobe Systems Inc., and widely used. PostScript is supported by many software packages and printers and is generally regarded as a *de facto* standard.

PostScript is available only from Adobe Systems, despite the fact that many of its details are publicly available, and are not protected as trade secrets. Adobe is able to maintain this position because it vigorously protects a key component of the technology – its method of storing and reproducing typefaces. As a result, competitors *can* produce (under licence) devices which understand the PostScript language, but the results that they produce are inferior to those of Adobe's own products.

Case 2: competitive products to single standard

Sometimes a single *de facto* standard appears from a competitive situation. The process is illustrated in Figure 4.2, where, of several innovative products designed for a particular task, only one emerges as the *de facto* standard.

Here, customers have made a choice between competing products. Sometimes this is based on technical features, but more often it is on issues such as price, availability, or quality of service from the supplier. After a period of time, a single product dominates the market and becomes the *de facto* standard.

The video cassette provides an example of this process. The VHS format has become predominant around the world, effectively pushing the competitive format, Betamax, out of the market.

Although Betamax was judged by most commentators to be technically superior, VHS succeeded because its originator, the Japanese Victor Company (JVC), made the format easier and cheaper to license than Sony's Betamax. Today, while continuing to support Betamax, Sony itself also markets VHS equipment and *both* VHS and Betamax have become public standards.

Another example exists within the personal computer market, where the IBM personal computer and its operating system, Microsoft's MS-DOS, both

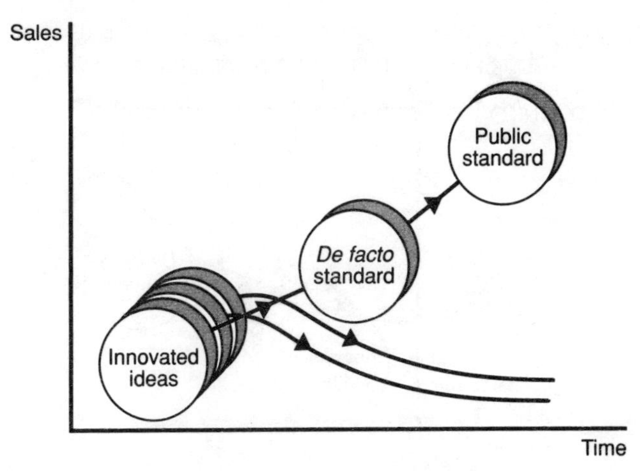

Figure 4.2 Competitive products to single standard

became *de facto* standards in the face of competition. Yet while the IBM PC has been widely copied, or 'cloned', the same is not true of MS-DOS. Microsoft, through a policy of selling in high volume at low prices, remains the sole supplier. PC hardware clones, on the other hand, appeared despite IBM's early efforts to suppress them.

IBM's tactic of attempting to protect the copyright in the computer's low-level control software—the basic input/output system, or BIOS—failed when competitors were able to reproduce the program's function without actually copying the individual instructions of IBM's program. Thus, a *de facto* standard came into being even though its originator was trying to prevent it.

Neither the IBM PC nor MS-DOS is yet the subject of a public standardization effort.

Case 3: several products to several standards

The market does not always decide between alternatives, particularly when a number of similar products are available. For example, the early promoters of 45 rpm and 33 rpm records were competitors, vying for total control of the market. They eventually realized that the public wanted both formats. This situation is illustrated in Figure 4.3, where, of many good ideas, several *de facto* standards emerge, leading to public standards that incorporate features of many of them.

An important example is provided by developments within AT&T's UNIX operating system for multi-user computers. By 1984, three major variants of UNIX had appeared within the marketplace, and were in widespread use. These were UNIX, as supplied by AT&T; BSD UNIX, as supplied by the Berkeley Software Distribution; UNIX as supplied by Microsoft and sold under the name Xenix. Although all of these derived from AT&T's software source code, they differed in detail. A public effort by the US IEEE took almost five years to agree on the public standard, which they called POSIX. This now incorporates features from both AT&T and Berkeley UNIX

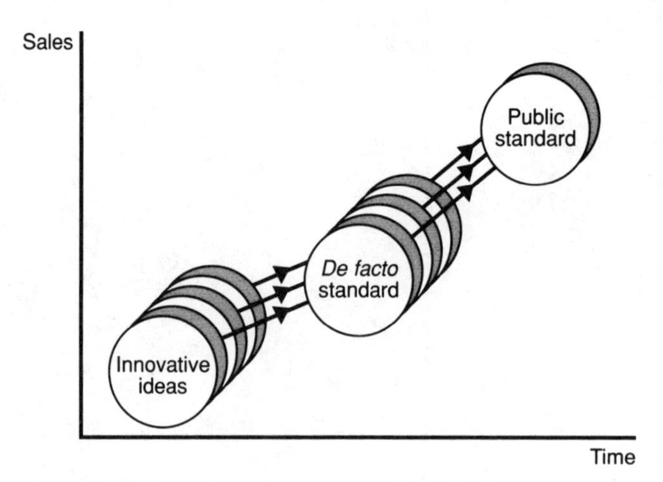

Figure 4.3 Several products to several standards

versions and is described in detail in Chapter 5. Since Microsoft did not play a large part in the formulation of the POSIX standard, no features specific to Xenix were incorporated into the POSIX standard.

Case 4: speculative development of standards

Sometimes a company will develop a technology with the speculative aim of creating a *de facto* standard. The result may subsequently be accepted by purchasers, or it may not. If successful, a public standard may ultimately result. This process is illustrated in Figure 4.4.

An example of this process is provided by the compact disc, developed jointly by Philips and Sony. After being heavily promoted and licensed to competitive suppliers, the format took off after an uncertain start, and has gone on to form the basis for a series of public standards.

This process does not always lead to a public standard. For example, video

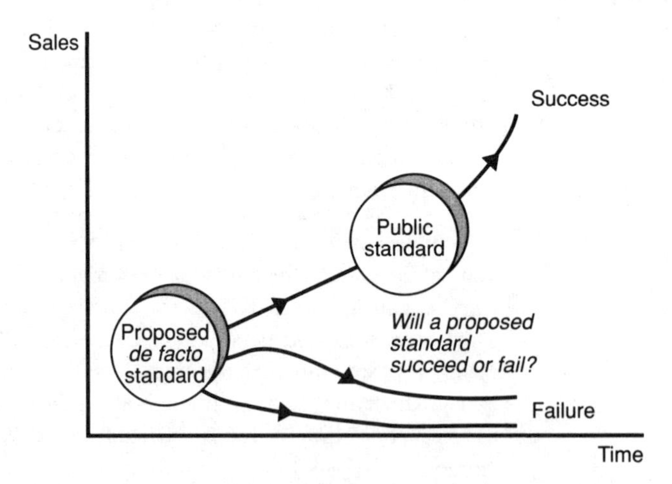

Figure 4.4 Speculative development of standards

discs, based on similar technology to that of video cassettes, have sold poorly in comparison.

In the computer industry, Sun Microsystems stands out as a speculative proposer of *de facto* standards. Sun's Network File System (NFS), a means of accessing files held on remote computers as if they were held locally, has been extremely successful. Widely licensed from Sun to other suppliers at no charge, it is now available for hundreds of types of machine, from personal to supercomputers. As a result, NFS is now being considered as a public standard by the appropriate bodies, starting with the IEEE.

NFS was introduced in 1985, when Sun Microsystems was just two years old. Sun's later proposals, the NeWS windowing system for visual displays, and the SPARC microprocessor architecture, have not achieved such widespread acceptance. This may be in part because the company's competitors are nervous of Sun's success and do not wish to contribute further to it. Indeed, one of the industry's group efforts to develop a major *de facto* standard, the X Window system (see Chapter 5 and Appendix 1), was largely funded by Sun's competitors, and provides an alternative to NeWS.

The X Window system has some features in common with a public standard, having been developed at a university (MIT) and funded by a consortium of computer companies. As a result it has moved rapidly into the public standards process.

Government bodies can also create *de facto* standards. An important example in the world of open systems can be seen in the Computer-Aided Acquisition and Logistics Support (CALS) project, a specific standardization effort being pursued by the US Department of Defense (DoD). This project has the purpose of 'defining standards which will allow the creation of a standard environment through the exchange of digital data between linked systems'. To this end, the DoD is specifying standards for data interchange, data management and access, and communications protocols. Where existing or emerging standards exist that satisfy the DoD requirements, these will be accepted within CALS; where they do not, they will be created.

In a similar way, a series of communications standards drawn up by the US Defense Advanced Research Project Agency (DARPA) have become very widely used. They include the TCP/IP (Transmission Control Protocol/Internet Protocol), which is widely used on local-area networks (see Chapter 6).

These US government communications standards take the form of a series of 'requests for comments' (RFCs) drawn up by US defence contractors. Because they are not the result of an open and formal process, these standards cannot be considered public. However, they differ from most *de facto* standards because there is no commercial interest behind them.

Setting public standards

In the previous sections, we have seen that *de facto* standards can appear from a variety of sources, and the same is true of public, or *de jure*, standards.

Like *de facto* standards, public standards are sometimes developed speculatively, rather than from market experience (see Figure 4.5). Such standards may be accepted or ignored by the marketplace. Many of the standards related to data communications have been created in this way. The most important example is that of the Open Systems Interconnect (OSI) data communications standards, which were developed by the International Standards Organization (ISO) with the intention of their becoming international standards.

Normally, public standards bodies are very wary of such 'invention', but OSI is considered a special case. Since all governments have a strong interest in data communications, the unusual decision was taken to standardize first, and then implement.

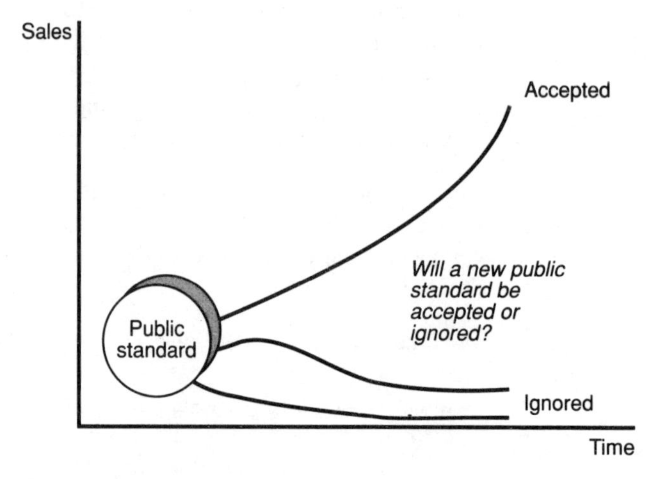

Figure 4.5 Setting standards

Over the years, the bodies which define public standards have settled into a stratified framework (see Figure 4.6).

Almost every country in the world supports a national body, usually an arm of its civil service, responsible for the formulation of public standards. Examples are the American National Standards Institute (ANSI) in the USA, the Deutsches Institut für Normung (DIN) in West Germany, and the Japanese Institute for Standards (JIS).

Public standards can cover anything from screw threads to software. Because of this, many national bodies delegate work on particular topics to certain professional associations. Known as *accredited bodies*, these organizations write standards on behalf of the national body.

The IEEE standard for operating system interfaces, POSIX, was handled in this way. ANSI delegated the production of the formal standard to the IEEE. On the other hand, ANSI itself has created standards for many computer languages, including C, COBOL and FORTRAN, without delegating to an accredited body.

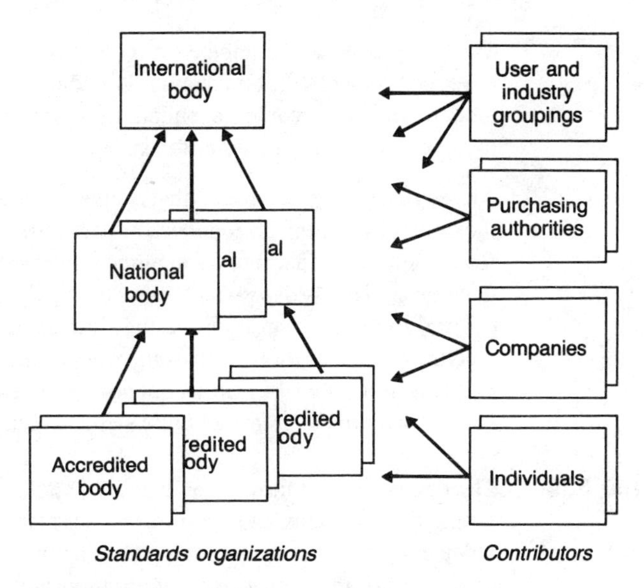

Figure 4.6 Contributors to standards organizations

International groups

The international bodies sit above the national standards bodies. In the field of information technology, the most important of these are the International Standards Organization (ISO), and the Comitée Consultatif International pour la Telégraphie et la Teléphonie (CCITT). ISO is concerned with standards in many areas, CCITT with those which relate to communications. Both organizations are headquartered in Switzerland, if for no other reason than to underline their international (neutral) status.

The open process

Formal public standards must be defined through an *open process*. This generally requires that all interested parties be allowed to participate and that such parties reach a high level of consensus before the standard is accepted.

Interested parties may be any or all of the following:

- *individuals*, who are assumed to represent only themselves;
- *companies* which want either to supply or to purchase standards-conforming products;
- *authorities* which determine purchasing policy for government or military organizations;
- *groups* of suppliers, of purchasers, or of both (examples are industry associations and user groups).

The policies of standards groups on the classes of participant that they allow vary. For example, the IEEE officially recognizes individuals only, and makes special provisions for group representation, while ANSI only allows representation at the company level, treating groups as if they were companies. In practice, many of the individuals who work on IEEE standards committees represent the views of their employer, rather than themselves.

The people who participate in public standardization efforts are almost always volunteers. Public standards bodies usually only employ administrators, relying on people seconded from industry to do the technical work involved in defining new standards.

Although helping to ensure that standards are drafted by those who know most about the subject, reliance on part-time voluntary labour is one reason for the long time that it takes to agree public standards. The volunteers participate in *working groups* which may meet only a few times a year. Monthly or quarterly mailings can improve the flow of information in the time between meetings, as can the increasing use of electronic mail, but substantive progress is limited by the frequency of meetings. Consequently, the open process required to create a public standard can take a long time.

Consensus and the open process

The open process follows legalistic rule books, which define membership and voting qualifications, the balloting process, and the handling of objections. In many cases, the rules governing particular processes specify a minimum delay of 30 days or more, in order that all interested parties have the chance to participate and respond. As an example of the procedures, those of the IEEE are outlined later in this chapter, and show the open process at work.

In general, consensus is reached by voting on successive drafts of a document, with a large majority in favour (sometimes even a unanimous vote) being required before a standard is accepted. Inevitably, participants must compromise in order to arrive at a document which, as well as being acceptable to themselves (or to the organizations that they represent), is acceptable to a sufficiently large number of the other participants.

It is not unusual for a standard to be several years in the writing. Even then, the draft standard which results may not be ratified if there are objections from those who are to vote on its acceptance. For example, ISO Draft Proposal 646, which describes international variants of the widely used ASCII character set, dates from 1983. It was only formally accepted as a standard in 1989, after the USSR dropped a long-standing objection to the inclusion of the US dollar-sign in all variants.

Classes of standards

As an aid to bringing standards into use as early as possible, three broad classes of standard are defined: the minimal standard, the compromise standard and the maximal standard (see Figure 4.7).

The minimal standard (Figure 4.7(a)) is a standard which embodies only those features about which the participants can easily agree. It avoids areas known to be controversial. When products exist which are competing as the basis for the standard, the minimal outcome is the intersection of the product definitions, i.e. it consists of those features that are common to all contenders.

An example of this type is the ANSI standard for database access, SQL. Many existing products conform to the SQL standard, despite the fact that implementations differ due to the inclusion of features which are outside the standard. If an application program uses features in one SQL product which

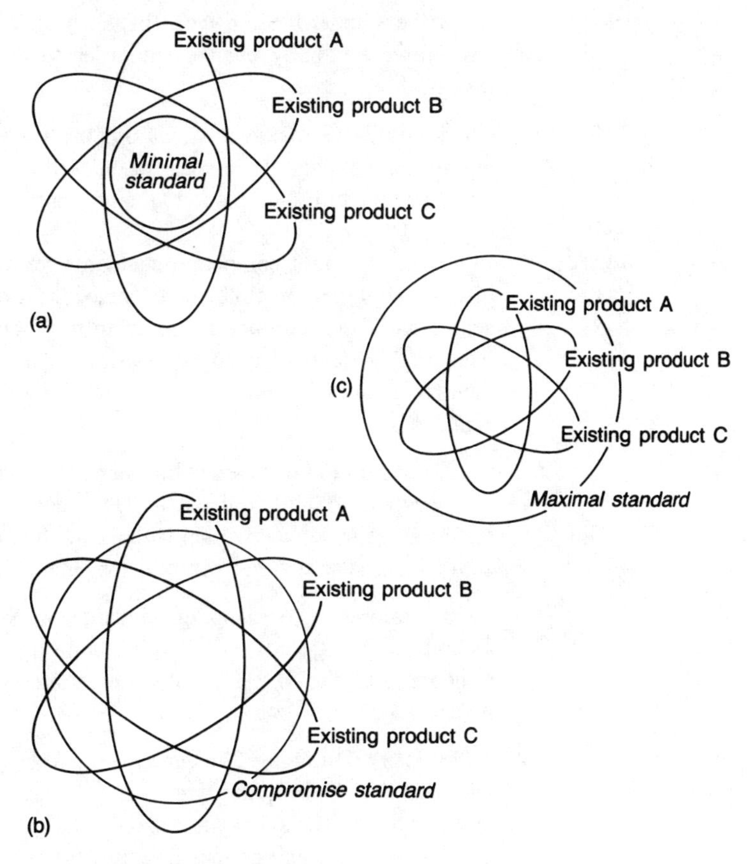

Figure 4.7 Classes of standard

are not specified by the standard, it is unlikely that the program can be run with another SQL product, even though both products may conform to the same (minimal) ANSI SQL standard.

Although the minimal standard is the easiest on which to reach consensus, it is the worst possible for the market. At best, a minimal standard can only be accepted as an interim step in the standards-setting process.

The compromise standard (Figure 4.7(b)) adds a few new features relative to existing implementation, typically because features found in some implementations are judged to be worth including in all systems. Completely new features are usually avoided. In this situation, most implementers will have to amend their products in order that they conform to the standard.

The IEEE POSIX standard is a compromise standard. Relative to AT&T's System V version of UNIX, it adds features from the Berkeley implementation; relative to Berkeley UNIX, it adds some AT&T System V features.

The maximal standard (Figure 4.7(c)) builds on existing practice, but produces a result more comprehensive than any single implementation in use when the standardization work began. Consequently, implementers usually have to make many changes in order to conform to the standard which results.

An example of a maximal standard is the 1988 ANSI standard for the C programming language. Five years in the writing, this specifies a language with many new features compared to the original language of 1983.

Overlapping standards

In order to prevent competing standards from defining the same concept in different and incompatible ways, one of the strongest rules underlying public standards is that conflict and overlap must be minimized. Thus the people who draft a standard must be aware of other public standards which affect their own, and must never redefine anything for which a description already exists.

Close cooperation between public standards-making organizations is essential if overlap is to be avoided. *De facto* standards seldom observe the same restriction, and so can overlap with each other, and with public standards, in a confusing and sometimes contradictory manner.

As an example of overlapping standards: AT&T's UNIX System V Interface Definition (SVID) covers much the same ground as the ANSI C language standard and the IEEE POSIX operating system interface standard (see Figure 4.8).

AT&T's SVID defines 236 functions, overlapping with POSIX in 52 cases, with ANSI C in 115, and with both in three. This situation results from the fact that the SVID is a *de facto* standard which predates both public standards by several years, and defines functionality which has yet to be formalized in the public forum.

Cooperative standards

Today's worldwide market for the products of information technology makes international standards more desirable than those conceived for narrower, local bases. While some public standards are drafted directly at the international

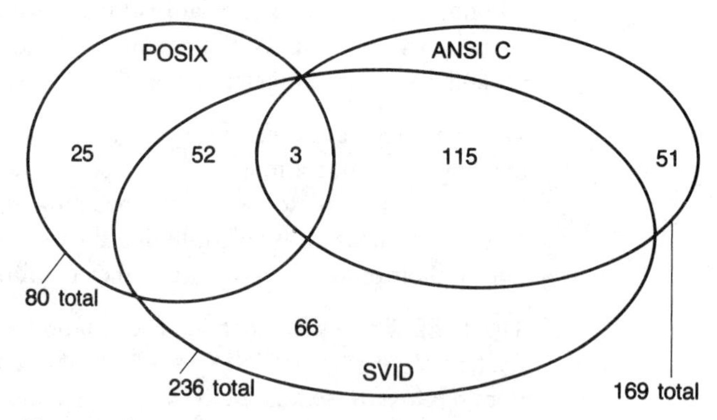

Figure 4.8 Overlapping standards

level (for example, the OSI family of networking standards), many start out at national level, or under the control of an accredited body. (See Figure 4.9.)

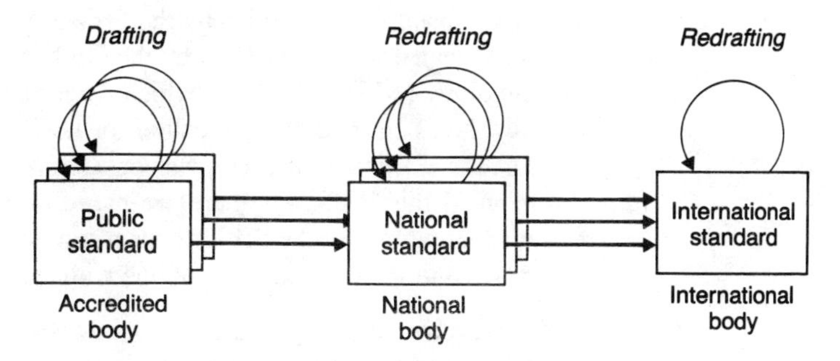

Figure 4.9 Evolution of standards

By providing advance information to a national standards body, and accepting advice on requirements in return, an accredited body can produce a standard which is immediately acceptable at a national level. An example here is the IEEE POSIX Operating System Interface standard. The IEEE cooperated with ANSI, with the result that the finished standard was accepted without change as a US national standard. (See Figure 4.10.)

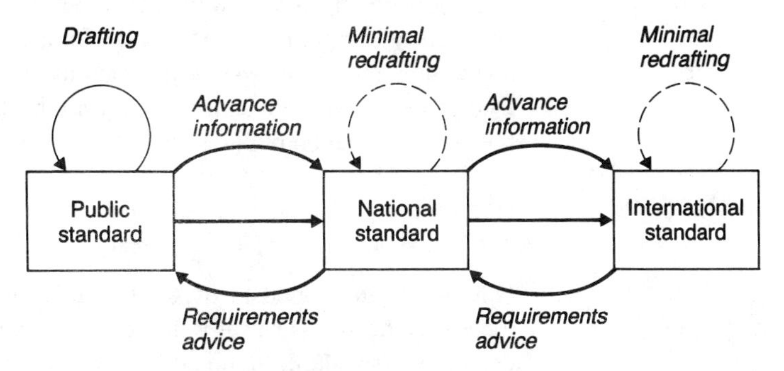

Figure 4.10 Standards making improved

Similarly, ANSI cooperated with ISO in the creation of the C standard, which consequently was able to become an international standard without changes. This was not the case for the original IEEE POSIX standard however. Several areas not covered in detail in the original US standard, but which are important in an international context, made it unacceptable to ISO without modification and subsequent delays.

Agreed standards

Delays in the appearance of standards are frustrating both to vendors, who want to implement conforming systems, and to purchasers who want assurance of conformance in the systems that they buy. Consequently, it is not unusual for both vendors and purchasers to decide to conform, or require conformance to, public standards which are still in draft form. In other words, they agree the acceptability of the standard before the full details are known.

For example, manufacturers supplied language compilers which promised conformance to the ANSI C standard—and which, in most cases, required revision when the approved standard finally appeared.

Another example is provided by the Federal Information Processing Standard (FIPS) created in August 1988 by the US National Institute of Standards and Technology (NIST), the body which determines US government purchasing policy. NIST needed a specification for UNIX-compatible operating systems before the IEEE was able to reach consensus on the POSIX standard. The first version of this FIPS was therefore based on number twelve of the thirteen drafts that it took for the IEEE to reach consensus. As a result, the FIPS had to be revised in 1989 to bring it into line with the final POSIX standard.

Most government departments must have policies that are demonstrably unbiased and so special care must be taken when they agree standards. If it is not, it may be challenged. An example of this occurred in the US in 1988 when the US Air Force required conformance to AT&T's SVID in a contract (AFCAC 921) worth almost $US1 bn. At the time, AT&T's SVID was the *de facto* standard for the UNIX operating system and its programming environment. Digital Equipment Corporation challenged the awarding of the contract to AT&T, on the basis that the specification of the system had been favourable to AT&T.

This contention failed in court when it was judged that the bid was sufficiently competitive, since SVID-compliant systems were available from many suppliers. However, organizations like NIST must continue to hold themselves above such charges. They do this by basing their specifications as much as possible on public standards, such as POSIX.

Options in standards

One of the problems with agreeing standards is that they may end up as minimal standards with optional features.

Optional features tend to make standards too general for practical procurement specifications. Usually introduced as a means of breaking deadlock in creating a standard, optional features usually offer one of three possibilities:

- *Alternative ways of implementing a feature.* As an example, the ANSI SQL standard allows two ways of reporting 'not found' errors through a special variable, SQL CODE.
- *Features which are not mandatory.* An example is the POSIX job control system, which allows running programs to be suspended and restarted by the user.
- *Ranges for valid values of variables.* An example is provided by the V22 modem standard, which allows momentary loss of the data-carrier to be tolerated for anything between zero and twenty milliseconds.

The core of a standard must be fixed between implementations, so that no matter who supplies a standards-conforming system, the buyer can be confident that the core will be provided. The same is not true of options since they may or may not be supplied.

Figure 4.11 shows four of the sixteen implementation possibilities for a standard which allows four distinct options. Four is a very small number in this context; most recent standards specify many more options. Obviously, this is an undesirable situation in the marketplace, since it quickly leads to very many variations in the possible products.

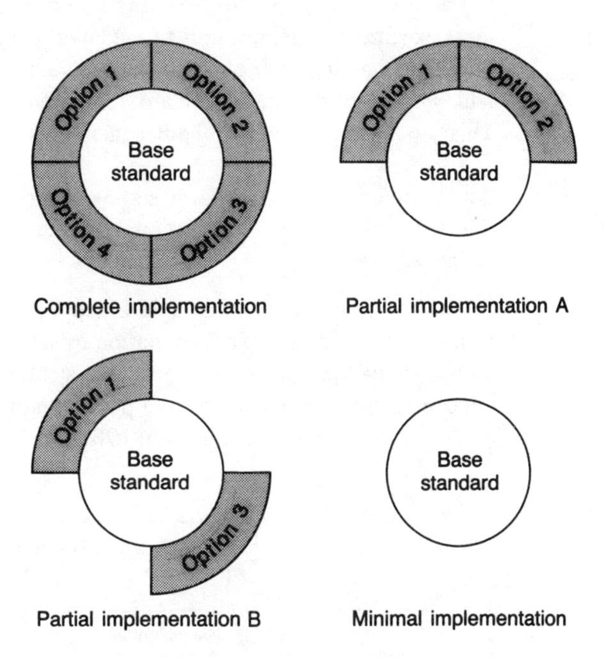

Figure 4.11 Implementation possibilities

In theory, such diversity allows a single standard to cover systems with a wide variety of capabilities. This allows buyers to specify only the options that they require for their application, without the need to pay for features that they do not need. In practice, the presence of options makes standards less useful. For example, if software is to be portable across as many implementations of a standard as possible, only core features can be used. The existence of options makes the writing of such programs difficult, since the options present in any one implementation must be known and avoided.

Options can also create problems for those who write the test procedures to check whether particular implementations conform to a standard. The more alternatives presented by optional features, the more difficult it is to write a test which copes with all the allowable combinations, but still correctly flags any implementation errors.

Because of the problems caused by options, purchasing authorities try to minimize their effects. In the three cases given above, an authority might remove them by stating:

- 'Such and such a feature must be provided, and must be implemented as follows . . .'
- 'Implementors must provide feature so and so in the following manner . . .'
- 'The only allowable value for such and such a quantity is . . .'

The result is that the purchasing specification is more rigorous than the public standard on which it is based. Purchasing authorities can often make expedient executive decisions, unrestricted by the requirements of an open process. Suppliers wanting their business then have no choice but to produce products which conform to public standards, as amended by the purchasing authority. This can make procurement bodies themselves important standards bodies.

The making of a public standard: the POSIX example

As an example of the standards setting process at work, we consider in detail the steps that led to the international standard, POSIX, for the interface to the operating system.

The US Institute of Electrical and Electronic Engineers (IEEE) has for many years been engaged in the formulation of standards. Over time, it has set up a well defined framework in which new standards may be produced. Initially concentrating on work relating to power engineering, the IEEE became concerned with computer standards in 1961.

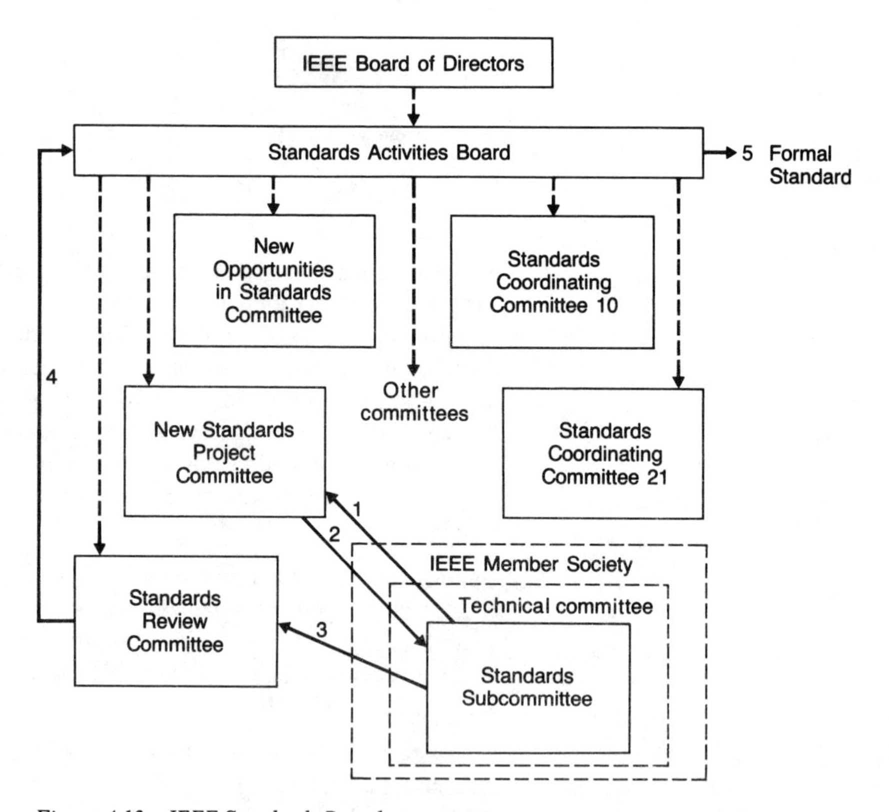

Figure 4.12 IEEE Standards Board

The simplified diagram (Figure 4.12) of the IEEE Standards Board shows a few of its key committees. Five other committees which report to the Board of Directors are not shown.

Overall control rests with the Standards Activities Board which, through its 'New Opportunities in Standards' Committee, looks for areas which warrant standards activities. With the aid of its Standards Coordinating Committees, the Standards Committee farms out the work of production to the IEEE's nine member societies. It is usually the societies themselves that identify the need for standards in their own areas of interest. For example, it was the Computer Society of the IEEE which, in 1985, proposed a standard based on the programming interface of the UNIX operating system.

When a need is identified, the society (or, strictly, the Standards Subcommittee of the Technical Committee of the society) puts a Project Authorization Request—a proposal—before the New Standards Project Committee (path 1 on the diagram).

The New Standards Project Committee considers the request, passing it up to the Standards Activities Board and its other committees. Because any overlap between public standards is discouraged, an important consideration is whether the proposed standards activity will conflict with other standards, generated either inside the IEEE, or by other bodies such as ANSI or ISO. Indeed, Standards Coordinating Committee 10 maintains a dictionary to cross-reference standards, and to ensure that all standards use the same terms when discussing similar concepts. Any conflicts must be resolved before the proposal is approved and handed back to the proposing society's standards subcommittee (path 2 on the diagram), along with a project number. Hence P1003, the project number allocated to the Computer Society's proposal, which ultimately became known as POSIX, was allocated here.

At this point, work starts, with the aim of producing a final standard to be approved first by a ballot of voting members of the standards subcommittee, then by the Standards Review Committee (path 3), and finally by the Standards Activities Board (path 4). Although often a formality, the last two stages form part of a system of checks and balances which exists to ensure that standards are properly and, as much as possible, promptly produced according to the rules governing the IEEE's open process. Only when a standard has passed through all stages does it emerge into the outside world (path 5).

On receiving authorization to proceed, the standards subcommittee forms a working group, which first appoints a chairperson, and then starts the work of drafting a standard.

A working group usually has three types of member:

- *Working Group Members*, who are active in the creation of the standard, and who can be expected to attend the quarterly meetings of the working group;
- *Balloting Group Members*, who, in addition to the duties of a working

group member, undertake to review new drafts of the standard, and to vote on their acceptability;

- *Correspondents*, who simply receive mailings of the group's working papers.

The IEEE will accept any individual, whether a member of the IEEE or not, to any of these classes of membership. Members are considered to represent themselves, rather than their employer, as they do on ANSI, or their country, as in ISO.

Working group members may be from either the users or the suppliers of technology, since public standards are developed to help both. Although the IEEE is based in the US, participants may be of any nationality whatsoever. Figure 4.13 shows the working group structure, both from the point of view of interests, and from the point of view of activity.

There are some organizations which could legitimately claim a right to participate in the standards-making process. The three industry groupings X/Open, Unix International and OSF, together with the UNIX user groups, Uniforum and Usenix, are examples. A special mechanism allows such organizations to participate as *parties of interest*, effectively giving them slightly more rights than individual members of the working group.

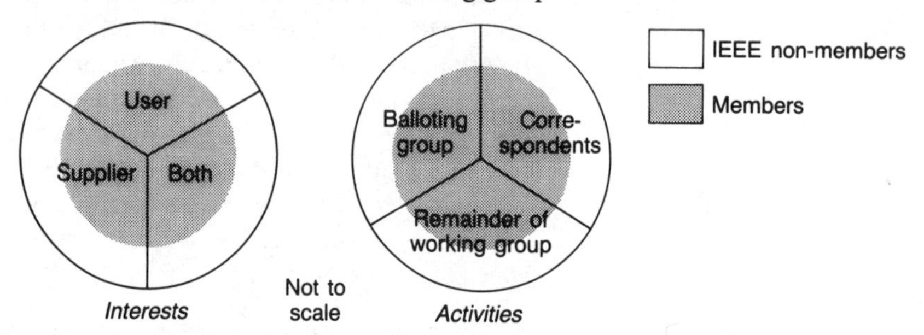

Figure 4.13 IEEE working group structure

A problem with the IEEE's Working Group structure could occur if a company tried to promote its special interests by allowing several of its employees to participate. If this seems to be the case, members of a Working Group can appeal to the Standards Review Committee, which has the power to change the structure of the working group if it sees fit. The Standards Review Committee can also step in to break a deadlock, if repeated ballots in a working group fail to obtain consensus. Although intervention of either type is rare, both have happened on occasion.

Any working group member, whether they are a member of the IEEE or not, can join the balloting group, but in order to be accepted as a standard, a draft must be accepted by at least 75 per cent of the IEEE members in its balloting group. In practice, this figure is considered marginal. Groups like to achieve acceptance by 90 per cent or more of the members.

The balloting group must balance the interests of suppliers and users. In order to ensure that neither has an overall majority, individuals are asked to declare their interest when they join the balloting group. Parties of interest may join balloting groups, but are given only a single vote.

Members of the balloting group must review each draft of a standard within 30 days of receipt. They can then vote in any one of the following three ways:

- accept without comment;
- accept with comments;
- reject with comments;

There are no other choices and voting is almost mandatory.

The timing of voting was determined many years ago when standards were simpler, and has remained unchanged since. In practice, it is almost impossible today to review the several hundred pages which make up the typical standard in only 30 days. Consequently, informal arrangements are usually made to share the work, with particular members of the group undertaking to review sections of the draft in detail.

A major purpose of the balloting procedure is to generate comments on ways in which the current draft might be improved, or on changes which must be made before a particular member of the balloting group would find it acceptable. Consequently, any negative vote must be accompanied by comment on the reason for rejection.

A ballot may generate a very large number of comments. Several thousand comments, ranging from notes on typographic errors and language to substantive technology issues, have been produced on occasion. The Working Group is bound to respond in writing to every one, and to attempt to resolve the comments which accompany negative votes.

This is not always easy; a change which turns one person's rejection into an acceptance may well prove unacceptable to several people who had previously accepted the draft. Sometimes, it is impossible to resolve a negative ballot. It is the task of the Working Group's chairperson to ensure that the number of outstanding negative votes is minimized, and that all substantive issues are properly addressed. To avoid rejection, the chairperson usually works to ensure that the contents of a draft are generally acceptable before it goes to ballot.

Ultimately, since the 75 per cent acceptance refers only to IEEE members, it is the IEEE members on the balloting group who determine the fate of a standard. However, because all comments, whether from IEEE members or not, are treated with equal weight, non-members have an incentive to join the balloting group and to shape standards through their comments.

The production and revision of a series of drafts may well take several years. A working group with a large task ahead of it may, early in the process, publish a 'Trial-Use' document. This term 'Trial-Use' is really a misnomer as the document is almost certainly too immature for practical use. Its purpose is to elicit early comment from all interested parties and therefore identify difficult or contentious issues.

Only when it is considered likely that a draft standard will easily surpass the 75 per cent criterion, and not excite comments which raise substantive new issues, is a formal vote for acceptance as an IEEE standard taken. On passing this ballot, and a final stage of vetting by the IEEE's committees, the standard becomes official, and is published.

Summary: the making of standards

In this chapter, we have examined the mechanisms, both formal and informal, that exist for the setting of standards.

There are two main mechanisms: *de facto*, in which standards emerge through use, and public, which arise from formal procedures within accredited bodies. Public standards are often formal descriptions arising from the evolution of *de facto* standards. While formal standards are preferred by most procurement agencies, the time it takes to agree them is sometimes unacceptably long. This can lead to incompatibilities, if users pre-empt the standards bodies work.

In order to reach agreement on standards among representatives with competitive and opposing interests, public standards bodies are often forced to produce a minimal definition of the standard, sometimes with options within it. This is of limited use in the IT industry, since it can restrict portability of software and conformance testing of products.

If all segments of the IT industry are to obtain the standards that they need in the time-frames that they need them, it will be necessary to develop mechanisms that enable users to specify standard which incorporate both public and *de facto* standards, and a means by which they can evolve with those standards, without jeopardizing their current investments.

CHAPTER 5 Standards for portability and scalability

In Chapter 3 we saw that standards for open systems need to address the issues of portability, scalability, interconnection and ultimately, full interoperability if some of the problems constraining the growth of the IT industry are to be resolved.

In this chapter we will review the current state of standards for open systems as they affect portability and scalability, including both those that exist today and those that are rapidly emerging. This will include a review of the leading products available for implementation of the standards.

The requirements of portability and scalability translate into a need to standardize the environment on a single computer, within which the application is expected to run. Figure 5.1 shows a graphical representation of this environment, based on the components of the levels for standardization as used in Chapter 3. The central role of the operating system software can be clearly seen.

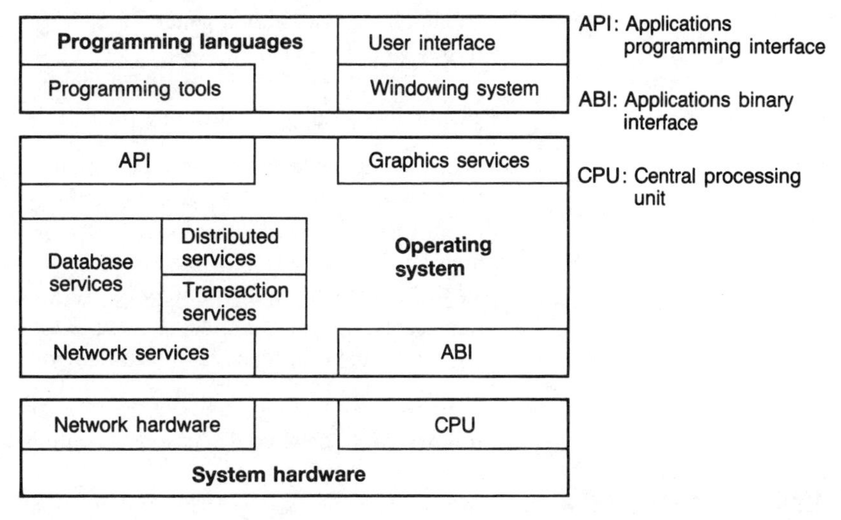

Figure 5.1 The application environment

Most standardization work on the operating system has been based on AT&T's UNIX operating system. This work has resulted in an initial public standard, POSIX, produced by the international standards bodies of the IEEE, ANSI and ISO, and so will be discussed in detail later in this chapter.

Standards work on distributed computing has so far concentrated on interconnection. This has been carried out within ISO, in cooperation with a number of industry groups, such as COS, OSITOP and SPAG, set up for that purpose. This work has resulted in the international Open System Interconnection (OSI) standard.

In addition to the OSI standards themselves, there are many associated standards required for the applications of open systems vital to major users. Prominent among these is EDI—electronic data interchange—and its many related technologies. Standards for interconnection and its applications are covered in detail in Chapter 6.

For interoperability, implying fully distributed computing, it will be necessary to merge the POSIX and OSI standards. This is happening on both the POSIX front, as the work expands to incorporate the OSI standards, and on the OSI front, as the work moves from the interconnection issues themselves, up to the applications. Both Chapters 5 and 6 will therefore touch on interoperability.

Standards for the application environment

For the full definition of open systems to be achieved, many components of the application software environment, including the operating system interface, need to be standardized. This is recognized by the standards groups, and is evident in the breadth of activity in their various subcommittees.

Although standardization generally takes place initially on the components as if they were independent, most of them have complex interdependencies. The most difficult part of the standards-setting process is in coordinating, and eventually merging, the different pieces of the standards work.

While standards for many of the components of the applications environment have been defined, at least in a preliminary form, in practice a sophisticated application will usually need facilities which have yet to be standardized. The high level of activity in standards bodies around the world means that more standards are continually emerging. But given the rapid advance of information technology, it is almost inevitable that the standardization process will lag behind the leading edge of technology.

A review of existing standards for the software components of Figure 5.1 follows. More detailed discussion is continued in the Appendices.

User interface (see also Appendix 2)

Today, the most visibly active and potentially controversial topic in the standardization area is that of the *graphical user interface* (GUI). Both users and suppliers would like to have a standard computer interface that is similar to that on the Apple Macintosh, with easy-to-use windows, menus and a keyboard/ mouse interface, and to have this interface common to most applications.

For UNIX systems, at least two commercial GUI products are competing for position as *de facto* standards. These are Open Look from AT&T, jointly developed with Sun Microsystems, and Motif, from the Open Software Foundation, based on technology from Hewlett Packard and Digital Equipment Corporation. Both these products have MIT's X Windows (see next section) software at their core, and both are widely available.

Within the public groups, an IEEE committee is dealing with the high-level user interface issues—those that impact applications software portability and concerns of 'look and feel'. This committee focused on the X Window work and coordinates with ISO groups working on lower levels of the software. Because X Windows software is independent of the operating system in use, this work is of wider scope than current IEEE POSIX work.

Within the user groups, the Uniforum Subcommittee on Usability and Graphics is defining interfaces for data interchange in the graphics area, and interfaces for window presentation managers. This work is also closely linked to, and influenced by, the X Windows work at MIT.

Windowing systems

The X Windows system, developed at the Massachusetts Institute of Technology (MIT), is designed to allow users simultaneous access to multiple applications from multiple computers across a network, displaying each application in a separate window on the screen. It is therefore potentially an important component of interoperability technologies.

X Windows was developed at MIT with funding from the US government together with contributions from the computer industry, notably from DEC and IBM. It is independent of the underlying operating system, and can therefore run on any system, open or proprietary. X Windows has been widely adopted throughout the computer industry and is a *de facto* standard.

The most important component of X Windows is its network support. X Windows is designed so that workstations or intelligent terminals containing the X Windows software (X-terminals) can access one or more computers on the network. X Windows can be integrated with graphical application interfaces that increase its ease of use. Because of this, it is built into many applications software products.

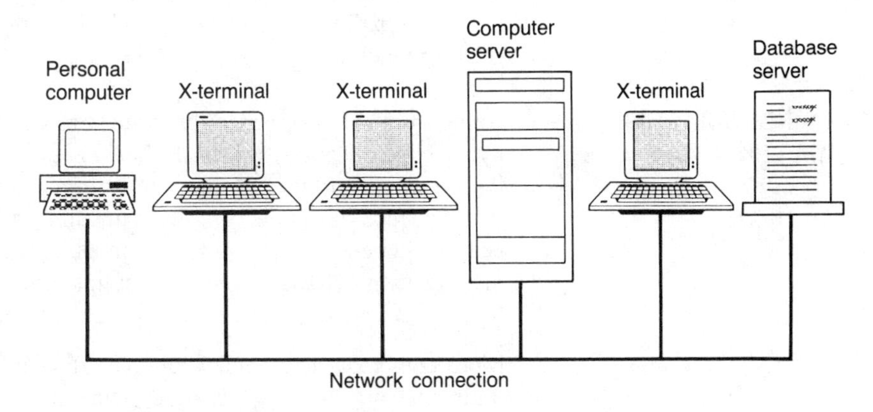

Figure 5.2 X-terminals on a network

Languages

Programming languages have long been subject to standardization, and it is possible to find many standards-conforming implementations of major languages, such as COBOL and FORTRAN, running under the UNIX, and other, operating systems. Closely related to standards for the operating system interface is the standard for the language C, in which the UNIX operating system is written.

The language C was informally described in a textbook, *Guide to C Programming* by Kernighen and Richie, in 1978. This remained the primary reference for users of the language for many years. In mid-1983, ANSI established the X3JIIC Language Standards Committee to develop a 'clean, consistent and unambiguous standard for the C programming language which codifies the common existing definition of C and promotes the portability of programs across C-language environments'.

This committee has maintained a policy of standardizing existing practices wherever a clear, unambiguous and sufficiently inclusive precedent is available. The standard for the C language which it produced was ratified by ANSI in 1989 but did not automatically become an ISO standard, because it failed to deal satisfactorily with internationalization issues.

Object-oriented languages are among the latest developments in programming, and consequently are some way from being standardized. The C++ language, a superset of the C language, was created at AT&T's Bell Laboratories to provide support for object-oriented programming techniques. Standardization for it has only just begun, even though several commercial implementations of C++, in its current definition, are available.

Programming tools and utilities

Hardware and software suppliers do not regard software programming tools as having a high priority for standardization, and seem content to wait for standards to be set within the public groups.

The IEEE is working on standards for programming interfaces to commonly used software tools and utilities. The concept is that it should be possible to write portable 'shell scripts' (batch command files) for application support or software installation and be confident that these will function properly on standards-conformant systems. Standards in this area are expected from the IEEE some time in 1990.

Applications programming interfaces

Application software generally requires more facilities than can be provided by a simple operating system. If the computer only supplies a simple operating system, then the application must implement the additional sub-systems that it needs. This is the case for many applications running on PCs with the MS-DOS operating system. For example, a database application must implement its own database in terms of simple file operations, and a communications program must provide low-level control for the hardware concerned.

More comprehensive applications environments, such as those supplied in traditional multi-user operating systems, offer the programmer more facilities, reducing the amount of work that the programmer must do. In open-system standards, the operating system is described in terms of the interface that it presents to the programmer, as are the additional sub-systems. A program written to work with a particular set of interfaces will then work on any system which supports (at least) that set.

How the support is provided is unimportant, as long as it is there. For example, if a system using the UNIX operating system could present the same application programming interface (API) as a PC running MS-DOS, then the UNIX system would be able to support programs written for MS-DOS. Similarly, if Digital's proprietary VMS operating system were to present an API which behaved like UNIX, then VMS could run applications written for that interface.

In the future, we can expect powerful and general operating systems, able to support a variety of APIs. For example, Mach (described below) has enough functionality to support both a conventional UNIX API (which ignores some of Mach's advanced features), and an API for Microsoft's OS/2 operating system.

UNIX itself cannot support an OS/2 API because it lacks support for a particular type of process control facility which is required in OS/2. On the other hand, OS/2 can support an API which conforms to the IEEE POSIX operating system interface standard, and so can Digital's VMS. Both Microsoft and DEC have announced their intentions to provide such an interface, in order that their systems conform to public standards.

Data management

The single most common reason for the purchase of computers is to manage data. Sometimes this data exists on a single machine, but increasingly, it is on a number of computers that need to communicate. A user often needs to access data on a computer in another location. Therefore the issues of data management are not independent of those of networking or of the user interface software.

All major suppliers of database management software are moving towards compliance with Structured Query Language (SQL), defined by an ANSI standard, which closely follows the query language of IBM's DB2 database product.

The ANSI SQL standard specifies two levels of compliance, level 1 and level 2. Level 1 defines a language subset which is too sparse to be useful, but with which it is easy to comply. To be practical, an implementation of SQL must include non-standard extensions beyond level 1. Level 2 is more comprehensive, and consequently more difficult to implement.

While the standard defines SQL as an interactive language, an extension describes how SQL statements may be embedded in programs written in other languages, specifically COBOL, FORTRAN, Pascal and PL/1, but not C. Since C is a very popular programming language, this is an obvious deficiency.

A specific and important area of data management that has yet to be standardized is that of transaction processing (TP). Applications which handle many operations involving several related, and possibly sensitive, data items quickly and safely can be considered a transaction processing system. Typical examples of such applications are high-volume order-processing systems, and banking control programs for networks of automatic teller machines. Transaction processing requires database management, reliability, rapid response times, and a high level of security.

The first TP systems were implemented using highly specialized software, often running on specially configured hardware. For example, most of the world's airline reservation systems use a specialized IBM mainframe operating system which was developed specifically for this application, and which is not used anywhere else.

Several suppliers have added proprietary extensions to their variants of the UNIX operating system to produce TP systems able to compete with the traditional TP environments. Because of the proprietary nature of the extensions, which differ considerably between suppliers, there is no standard yet for TP in an open environment, and so the potential advantages of portability and ease of system integration are so far not available.

For many years, the UNIX operating system as supplied by AT&T was deficient in the features which are necessary for reliable transaction processing. Examples are: the ability for one user to lock parts of a file, and so prevent other users from accessing it during critical operations; a mechanism to ensure that data has been written safely to disk before an operation can proceed; the capability of handling a large number of files simultaneously from a single job.

Recent implementations of UNIX from AT&T provide these and the environment defined by the IEEE POSIX standard allows for them. Implementation of TP applications on open systems is therefore becoming a commercial reality.

In the spring of 1989, work started on a public standard for TP within the IEEE POSIX committees, but results are not expected in the near future.

Security

As computerized information systems come into widespread use in corporations and government agencies, security is an increasingly important issue for almost all users.

All modern multi-user operating systems offer some degree of security. AT&T's UNIX is no exception to this, and, when properly administered, can be trusted to offer users reasonable privacy and protection of data from unauthorized access. But UNIX was not designed to protect highly sensitive information in hostile environments. If it is to do so, additional features are required. Adding these is likely to affect the work in many of the POSIX subcommittees.

In 1983, the US National Computer Security Center produced a document, *Trusted Computer System Evaluation Criteria* (TCSEC), popularly known as

'The Orange Book', which defined a number of levels of security for isolated systems. The TCSEC became a US Department of Defense standard in 1985. Implementors offering security features usually specify conformance to a level defined in the Orange Book. Security for networked systems is the subject of a later guide.

The IEEE Security Subcommittee is defining interfaces and structures for POSIX-conformant systems so that they can be built to comply with the DoD specifications. Meanwhile, AT&T's latest release of UNIX, System V Release 4.1, is designed to exceed level B2 of the TCSEC definitions.

Networking

Note: This subject is covered in detail in Chapter 6. It is therefore only reviewed briefly here.

Networking technology can be classified as transparent and non-transparent. With non-transparent networking, the user must be aware of the location of the data he wishes to access. The non-transparent network provides mechanisms for accessing remote data, but the method is different from that used to access locally held data. With transparent networking, the same method for data access is used, independent of the location of the data, and the user does not need to know where the data is located.

Sun Microsystems' Network File System (NFS) is the *de facto* standard providing transparent networking. AT&T's solution, Remote File Sharing (RFS), is considered better than NFS in provisions for security, reliability, recovery and network loading, but has not been widely adopted by other suppliers.

The *de facto* standard for network communications in the US is TCP/IP, the communications standard defined by the US Department of Defense. TCP/IP users, particularly those in Europe, now require systems that provide a migration path to the ISO OSI standard. Although there is much activity in this area, few practical communications products based on the OSI standard protocols have yet appeared.

The IEEE Networking Subcommittee is working on the networking tools and functions necessary for interoperability, specifically interprocess communication, transaction processing, remote procedure calls and file sharing. It is working within the OSI standards, as well as with TCP/IP.

Graphic services

A *de facto* standard for graphic services has emerged in the form of PostScript, a page description language (PDL) from Adobe Systems Inc. This allows text and graphics to be created and laid out on paper with a high degree of flexibility, using black-and-white printing, grey scales, and colour. PostScript has been widely adopted, particularly by manufacturers of laser printers used by desktop publishing packages, such as those found on the Apple Macintosh. A variant, Display PostScript, brings the same capabilities to computer screens.

While Apple and Microsoft provide simpler alternatives to Display PostScript, Sun Microsystems provides one which is arguably more powerful.

Sun's Network-extensible Window System (NeWS) provides full PostScript facilities in screen windows, remotely accessible over a network. Although AT&T's UNIX System V Release 4 includes the source code for NeWS, it is not clear how many of AT&T's customers will incorporate it into their UNIX-based products. Without that, NeWS will not become a *de facto* standard.

Multi-processing and the Mach operating system

The UNIX operating system has been in existence for over twenty years. As with any popular and long-lived program, it has been continually enhanced, and amendments have been made to the original source code in an increasingly disorganized manner. As a result, the software is difficult to maintain, and is suspected of harbouring hidden problems, particularly in the area of security.

Because the UNIX operating system expects to be in complete control of a single computer processor, whether micro, mini or mainframe, it is difficult to amend the source code so as to support new multi-processor architectures, i.e. computers where a number of processors are able to share the load and so (it is hoped) get more jobs done per unit of time.

AT&T, provider of the source code on which all commercial implementations of UNIX are currently based, is aware of its shortcomings, and is rewriting UNIX to address them. Results are not expected until the 1990s. But there is already a rewritten alternative, in the form of an operating system called Mach.

Developed at Carnegie-Mellon University (CMU) with funding from the US Department of Defense, Mach is designed to cater for computers providing anything from a single processor up to hundreds of processors. Object-oriented programming techniques are used, making the operating system modular, and, it is hoped, easy to maintain. Flexible methods of communication between and within running jobs make it irrelevant whether a computer has one or several processors. Mach is compatible with the Berkeley (BSD) variant of the UNIX operating system, giving it access to a wide base of existing technical and research software.

Mach will soon conform to the IEEE POSIX standard for operating system interfaces. Since the creators of Mach contribute to public standards efforts, it is likely that techniques for multi-processor support derived from Mach will eventually become public standards.

Although it has been implemented on many different types of computer, Mach is not yet available commercially. However, since it is the core of OSF/1, the operating system to be offered commercially by the Open Software Foundation, this will soon change.

Standards for hardware peripherals

One of the most neglected, and to the user, irritating areas in the computer industry is that due to the lack of standards for interfaces to hardware peripherals such as printers, terminals, plotters and storage devices. Contrary to most users' expectations, it is not possible to 'mix and match' peripherals, using printers and terminals interchangeably, without extra work.

Until recently, the only standards that have existed have been *de facto*, created by manufacturers such as IBM, with its 3270 terminals, and DEC with its VT100 terminal. Other manufacturers have produced terminals that are clones of these or have the ability to emulate them.

The increasing popularity of UNIX-based computers, which have the ability to support a wide variety of terminals and printers on the same system, has focused attention on the lack of standardization in peripherals. Although UNIX can handle the majority of the issues, doing so causes an unnecessary burden on the processor. Setting up the system, and providing technical support for it, can also put a heavy load on the supplier. In the case of a software problem experienced by a user, the technician not only has to understand the characteristics of the base hardware, the operating system, and the application software, but must also be concerned with which particular peripheral is demonstrating the problem.

In the case of new storage devices, such as CD-ROM, the problem of standardization has been dealt with early, with companies such as Microsoft and Lotus, contributing heavily to the standardization work.

Future needs already show requirements to integrate voice, images and video information with textual data. These will require standardization of new peripheral device interfaces if they are to merge into the rest of the open systems technology. While it is too optimistic to expect retrospective standards for peripherals already in use, we can expect a rationalization in this respect. For example, market forces are already demanding either Hewlett Packard laserjet or Epsom compatibility for printers, and Hayes compatibility for modems.

Internationalization

Users of computers speak many different languages and conform to different cultural conventions and business practices. It is therefore important that standard systems are capable of supporting a range of languages and cultural environments. In many cases, a strong requirement also exists to cope with these variations on the same system. An example is within the administration of the European Community, where it is often necessary to produce documents that incorporate several languages.

To date, the UNIX operating system and systems derived from it, have been based on the ASCII 7-bit coded character set and American-English. There are usually no facilities for dealing with other coded character sets, nor for supporting different languages and conventions. Those that exist are proprietary and therefore not portable.

The requirement for effective mixed language working brings with it the need for coded character sets larger than can be accommodated by 7-bit characters, as does the requirement to support the more complex languages. At the same time there is a trade-off between the ability to handle larger coded character sets, and the inefficiency that can result in the amount of storage required to hold the data. For most European requirements an 8-bit system provides the correct balance. For the major Eastern languages (such as Chinese and

Japanese) a 16-bit or larger system is necessary, even to support a single language.

To satisfy these primary requirements, enhancements must be made to provide full data transparency to applications, allowing flexibility in the choice of coded character sets employed. Additionally, the system must allow program messages (both input and output) to be handled in the native language of each user, as well as providing culturally dependent data items, such as date formats and currency symbols, to be used.

Most current implementations of the IEEE POSIX standard deal only with the 7-bit US ASCII character set. The Uniforum Subcommittee on Internationalization has pioneered much work to extend the POSIX standard to support European, Asian and other natural languages. This subcommittee works closely with the ANSI C language committee, with the Internationalization Working Group of X/Open and with the various POSIX subcommittees of the IEEE.

Because the issues of internationalization span all areas of standardization, there is an urgent need to incorporate internationalization standards into all other standards. Unfortunately, perhaps because so much standard activity takes place within the US, internationalization is sometimes not given the priority it requires. This was evidenced when the ANSI-ratified standard for the C language proved unacceptable to the international body, ISO.

Standards for the operating system

Having looked briefly at standards for each of the components of the application environment shown in Figure 5.1, we move on to look at the situation that prevails for the operating system itself. However, it must be remembered that the meaning of 'operating system' is itself evolving continually to include ever more of the components already discussed.

As is apparent from the discussion so far, the key ingredient in much of the open-system technology is based on the UNIX operating system, developed and supplied by AT&T. For this reason, it is important to understand at least a little of the historical development of UNIX.

History of the UNIX operating system

The UNIX operating system was written within AT&T Bell Laboratories by Ken Thompson and Dennis Ritchie in 1969. It was developed on a Digital Equipment Corporation PDP-7 computer, and written in a specially developed language, which Thompson and Ritchie called 'C'.

Today, C is the most popular systems programming language in the world and UNIX the single most widely used multi-user operating system. Both are supplied on a wide variety of computer systems, from portables to mainframes, from every major computer manufacturer in the world.

Around 1974, the UNIX operating system began to appear in university departments in the US and other countries, but because of the regulated monopoly status of AT&T at that time, it did not become a commercial product until some years later.

In 1977 the UNIX source code was made commercially available by AT&T, and newly formed software organizations began to offer it on computers such as the DEC PDP-11. One of these, Microsoft Inc., supplied a version which it called XENIX because it was unable to license the name UNIX from AT&T. XENIX, renamed UNIX in 1988 under license from AT&T, is now supplied principally through Microsoft licensee, the Santa Cruz Operation (SCO).

SCO has enhanced its XENIX/UNIX product considerably over the years, concentrating on the IBM PC and compatible market for computers using Intel microprocessors. SCO's product is supplied by many of the hardware manufacturers under license to SCO, and provides a multi-user alternative to single-user PC/MS-DOS systems. Today, more than half the worldwide installed base of UNIX systems, by units, is based on SCO UNIX.

AT&T's development of UNIX continued, with Version 7 becoming the basis of the University of California's Berkeley Software Distribution (BSD) implementation, widely used in academic institutions. While AT&T's own development continued, resulting in the current standard, UNIX System V, the BSD product line, which specialized in distributed computing and networking enhancements, diverged from it.

Some hardware manufacturers, wishing to offer products based on UNIX, licensed the operating system directly from AT&T. Prominent among these were IBM, Unisys, and Amdahl. Others, such as Sun Microsystems, DEC and Hewlett-Packard, took UNIX from AT&T, but incorporated many of the BSD enhancements.

All the computer manufacturers with UNIX source code under licence from AT&T added 'enhancements' to it. Sometimes these were no more than corrections to the code. Others added additional features required for applications such as security, transaction processing or real-time applications. Some added features in an attempt to provide competitive differentiation.

The result of these activities, by the software suppliers as well as the hardware manufacturers, was an explosion in the variants of the UNIX operating system as supplied to the market. By the mid-1980s, there were more than *100 versions* in active use. However, there were only three dominant versions—AT&T's System V, Microsoft's (SCO's) XENIX and Berkeley's BSD. All the rest were relatively minor variants of these three.

During 1988, AT&T, Microsoft, SCO, and Interactive Systems Inc. (a UNIX software developer) cooperated to converge the XENIX and System V versions of UNIX. This resulted in a UNIX System V product for the PC UNIX market that was able to run all the existing applications software written for XENIX. SCO named their version of this 'UNIX V/386', to denote that it was the standard UNIX V software, targeted to machines using the Intel 80386 microprocessor.

In 1989 AT&T released a new version of UNIX, System V.4. This incorporates features from BSD UNIX and Sun Microsystems's enhancements to it. By merging the three main versions of UNIX, AT&T has in theory removed fragmentation from the UNIX product lines. However, this will only be the case in practice if the hardware manufacturers refrain from making their own, proprietary enhancements to it.

Meanwhile, BSD UNIX continues to be developed, and is expected to remain at the leading edge of technology. Some manufacturers are expected to evolve enhancements to UNIX not currently envisaged in the standard product. Those additions which turn out to be of sufficient importance and of general use will eventually be incorporated back into new UNIX versions. Thus a continual process of divergence from and convergence to a single version of UNIX can be expected. This is the nature of standards in an evolving technology.

Standards for the operating system interface

While UNIX has emerged as a *de facto* standard for the operating system for multi-user computers, it is not a satisfactory basis for the definition of a public standard. UNIX is a proprietary product, owned by its developer, AT&T. However, its licensees, the other computer manufacturers, are competitors to AT&T's own computer business.

It is not realistic to expect a major computer manufacturer to be dependent on a competitor for development, licensing and supply of a strategically important product, particularly now, when the stakes in the open-systems market are very high.

For this reason, definition of the standard which will allow portability of applications software across machines has centred *on the interface to the operating system*, rather than the operating system itself.

The role of Uniforum (/usr/group)

The formal process of standardization of an *interface* for a portable operating system based on UNIX began in 1981, when the International UNIX User Group, Uniforum (previously known as /usr/group), formed its Standards Committee with the brief '*to define an interface to a portable operating system, based on UNIX, that will allow portability of applications across computers from any system that supports it.*'

Uniforum is a non-profit trade association dedicated to the promotion of UNIX and UNIX-like operating systems through the exchange of information and the cooperative efforts of its individual members and corporate sponsors. The group was founded in 1980 and incorporated in 1981. Today, it represents the interests of many thousands of vendors and users of UNIX worldwide.

It is ironic that the group that calls itself the 'UNIX User Group' was the organization that first recognized the importance of defining a standard interface to the operating system. Such a standard clearly allows other operating systems to be built, or modified, to meet the interface standard.

Over the years 1981–84, the definition of the standard interface evolved continually, with inputs coming from many sources. These represented all the components of the computer industry, and cooperation was universal.

In 1984, the proposed /usr/group standard for a portable operating system interface was adopted by the membership of the group. At that time, the proposed standard was transferred to the formal standardization procedures of the IEEE, with the intention of all the participants that it should rapidly become an international standard.

The Uniforum Committee responsible for the work on the operating system interface was formally disbanded, although many of the people involved in it moved on to the IEEE Committee. Meanwhile, the Uniforum Technical Committee initiated subcommittees to work in other areas deemed relevant to the application environment. These included internationalization, security, transaction processing, distributed file systems, network interface, real time, supercomputing, graphics and performance measurement (see Figure 5.3).

C++ Subcommittee

Distributed File Systems Subcommittee

Internationalization Subcommittee

Network Interface Subcommittee

Performance Measurement Subcommittee

Real-time Subcommittee

Security Subcommittee

Supercomputing Committee

Transaction Processing Subcommittee

Usability/Graphics Subcommittee

Figure 5.3 Subcommittees of the Uniforum Technical Committee (July 1989)

As the work in each of these Uniforum Technical subcommittees matures and consensus on a preliminary definition of the standard is reached, each moves across to the relevant committee in the IEEE standardization process.

The POSIX standard

In 1985, the IEEE formed its P1003 Standards Committee to work on 'Portable Operating Systems for Computer Environments'. Its intention was to turn the results of the Uniforum work into an international ISO standard.

To avoid using AT&T's proprietary name, UNIX, the IEEE Committee adopted the name 'POSIX', short for 'Portable Operating System based on UNIX', for its proposed standard. This IEEE Committee, now known as the 'POSIX Committee' first produced a 'Trial Use Standard P1003.1' in 1986, having taken a disappointingly long time to do so, in many people's opinion. In part, however, the time taken was a measure of the importance of the issues to many of the contributors.

At the end of 1987, Draft 12 of the 'full-use' POSIX standard for P1003.1 appeared. Draft 13, which was produced in 1988, became the *full-use standard* and was subsequently adopted, first by ANSI, and then by ISO.

As long as the standards work is in the proposal stage, 'P' remains in front of the name of the Committee. When the standard is adopted, the 'P' is dropped. Thus this first POSIX standard is now the 1003.1 international standard. All products targeted to the open-systems market are moving towards conformance with ISO 1003.1.

The POSIX committees have expanded in much the same way as those of Uniforum, and the IEEE now has subcommittees working in many components of the application software environment (see Figure 5.3). For ease of identification, each of these receives a decimal qualifier, to distinguish them one from the other, while retaining their immediate identification as part of the POSIX work. Thus, P1003.1 is the first of the POSIX subcommittees, and its work covers the interface between the application program and the operating system. Other components of the application environment are dealt with in the other P1003.n subcommittees.

P1003.0	**The POSIX Guide Project** Describes how the POSIX standards work together and how they relate to standards from other bodies.
P1003.1	**System services interface** Defines standard interfaces for the most basic levels of the UNIX operating system.
P1003.2	**Shell and tools (and user utilities)** Defines standard programmatic interfaces to the common UNIX tools.
P1003.3	**Testing and verification** Defines a testing standard for the 1003.1 standard so conformance can be verified.
P1003.4	**Real-time extensions** Defines an interface into 1003.1 for applications with real-time concern.
P1003.5	**Ada language binding** Defines the POSIX standards in terms of the Ada language.
P1003.6	**Security extensions** Defines interfaces and structures for POSIX for compliance with DOD's Trusted Computer Systems Evaluation Criteria.
P1003.7	**System administration** Charter is to define a standard set of system administrative tools.
P1003.8	**Networking extensions** Working within the confines of OSI protocols as well as TCP/IP, the scope of the group includes all networking tools and functions.
P1003.9	**FORTRAN language binding** Defines the POSIX standards in terms of the ANSI FORTRAN language.
P1003.10	**Supercomputing** Addresses big machine operating system requirements.
P1003.11	**Transaction processing** Addresses the issues of rapid processing of multiple transactions.

Figure 5.4 IEEE 1003.n Subcommittees for POSIX standards (April 1989)

Each of the components of the application software environment is dealt with by a P1003.n committee. The number of these will expand with time as technology advances. P1003.0 has the task of defining the totality of components that need to be standardized.

The name 'POSIX', which was originally used for the standard emerging from the 1003.1 Committee, is now often used to describe the 1003.n committees collectively. Thus 'POSIX' may refer *either* to the standard for the system service interface *or* to the entire application environment.

In general, preliminary work on standardization in new areas is done within the Uniforum committees. Only when it appears that consensus is emerging on the importance of the area and on a preliminary view of the standard is the work taken into the IEEE Committee. As in the case of 1003.1, most of the participants also move into the IEEE Committee and continue the work.

A diagrammatic history of the POSIX standard is shown in Figure 5.5.

Comment from a participant in the standards process

Although it can appear to the outsider that the standards processes are slow and inefficient, and that there is much duplication of effort in the committees, this is not felt to be the case by the active participants. Evidence of this is provided by Dr Brian Boyle, a long-term donor of time and energy both to the Uniforum and the IEEE Committees, and a leading authority on Internationalization in particular.

> As an active participant in the oldest and the most recently formed groups and as an original participant in the /usr/group and IEEE standards bodies, I have had the opportunity to observe the standards process in action over the years.
>
> Apparent disadvantages of the ponderous multi-organization structure offer unforeseen benefits, the Technical Advisory Committees, for example, acting as a vehicle for companies to offer valuable techniques and programs as less formal 'recommendations' which are carefully 'sanitized' of any vendor-specific biases before submission to the IEEE 1003, ANSI X3J11, ISO or other technical committees. 'Compartmentalization' permits the analysis, discussion and creation in small working groups that could never succeed in the endless debate of large groups better suited to *editing* than *synthesis*.
>
> I seriously doubt that dissent will ever truly cease or the process be smooth, but I perceive that as one of the 'biological' strengths that should make the standard and the UNIX operating system a viable force into the next century. As the Dean of the UCLA Computer Science Department remarked at a 1985 Conference of Supercomputing: 'I don't know how they will be programming scientific applications in the year 2000 but I'll bet the language will be called FORTRAN.' Similarly, while there's no safe bet as to the standard operating system of the year 2000, it is reasonable to expect it will be called UNIX.
>
> Dr Brian Boyle, Novon Research (1987)

Products for implementation of POSIX

The definition of POSIX and its acceptance as an international standard enables any company to design new, or modify existing, products to meet the

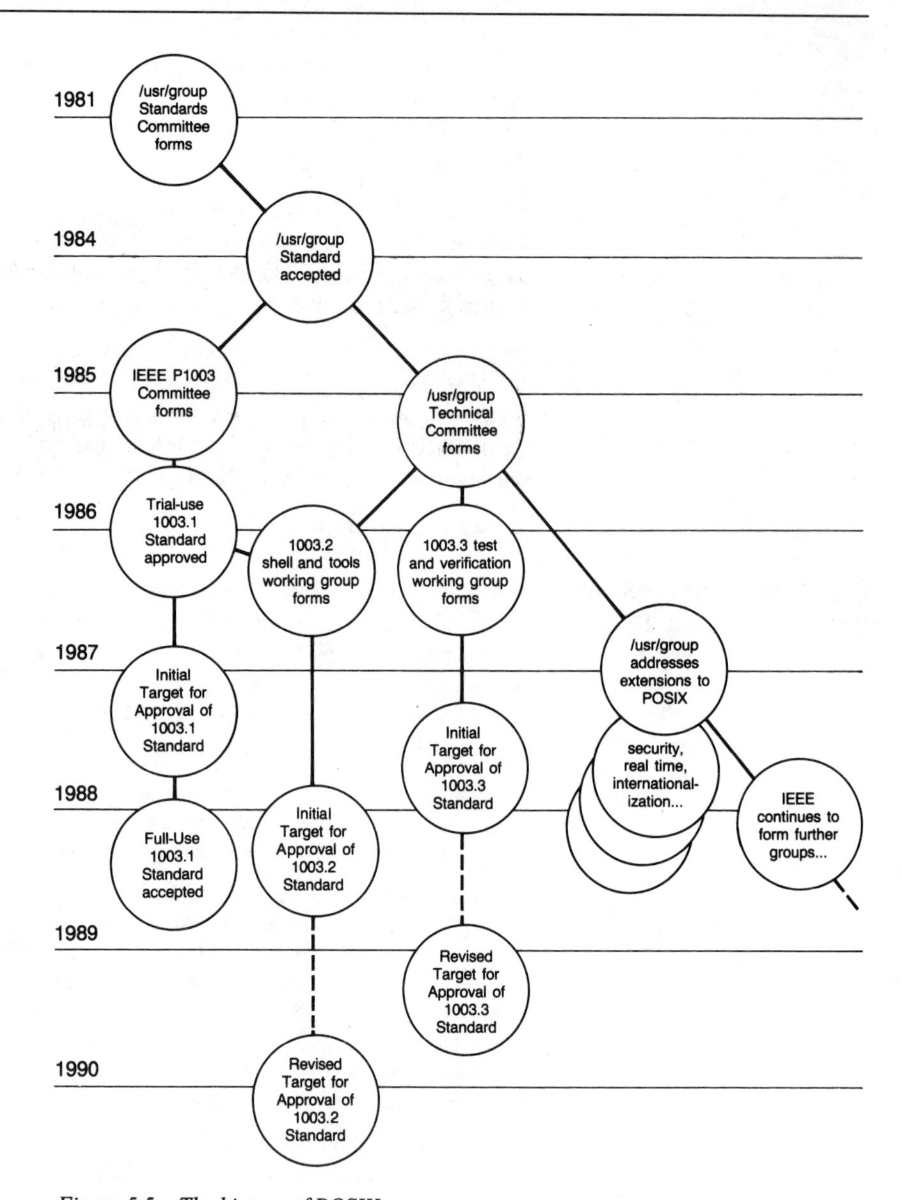

Figure 5.5 The history of POSIX

standard. This can cause confusion in the marketplace, particularly when there is a *de facto* standard product in existence, on which the standard has been based.

In the case of POSIX, there was such a *de facto* standard in use, AT&T's Unix, on which the POSIX standards work was based. As POSIX evolved, AT&T evolved UNIX to conform with it. At the same time, AT&T is developing UNIX in advance of the standards-setting process.

Other companies can be expected to produce products that meet the POSIX standard and which may be competitive with UNIX. Among these are:

OSF/1, the planned product from the Open Software Foundation (OSF); AIX, IBM's own internally developed UNIX variant; and VMS, DEC's proprietary operating system, which DEC has stated will be brought to POSIX conformance. Only the first of these can be expected to be made available under licence to other manufacturers.

In the sections that follow, we look at the current state of development of AT&T's UNIX operating system product, and at the proposals for OSF's OSF/1.

AT&T's UNIX System V

AT&T's formal document, the UNIX *System V Interface Definition* (*SVID*), introduced in 1985, sought to specify 'an operating system environment that allows users to create applications software that is independent of any particular computer hardware'.

Closely coordinating first with the development of the /usr/group standard, and later with that of the POSIX standard, AT&T itself has continually advanced development of its UNIX operating system. While doing this, AT&T promised '*to offer a solid and fixed target for application software developers for now and for the foreseeable future*'. One of its design objectives is to offer upward compatibility as new versions are produced. This means that applications software written for the old version of UNIX are guaranteed to run on the new.

In late 1989, AT&T released UNIX V.4, the UNIX version that merges XENIX and BSD UNIX with standard POSIX-conforming UNIX System V (see Figure 5.6). This product incorporates many enhancements, such as:

- support for current *de facto* networking standards (TCP/IP, RFS and NFS);
- software (Virtual File System Architecture) for handling very large amounts of data, possibly distributed over networks;
- improvements to the software for handling a variety of peripheral equipment from various suppliers (Device Driver Interface Specifications);
- specifications (Application Binary Interfaces) that will allow manufacturers using specific microprocessors to design systems that are completely compatible;
- software (Open Look Application Programmer Interface) to improve the user interface to the applications.

The implications of each of these is described later in this chapter.

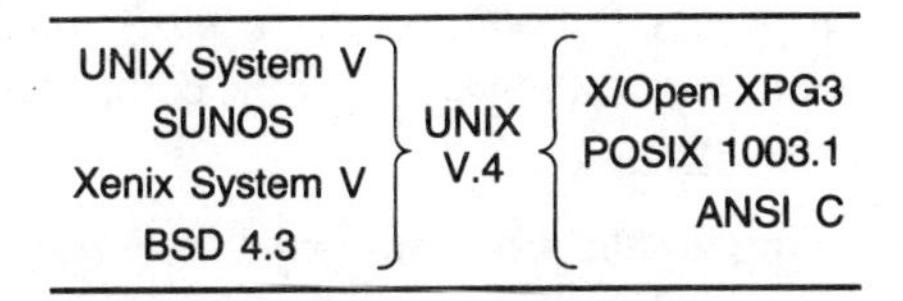

Figure 5.6 Convergence of UNIX versions

It is important to understand that not only is AT&T active in efforts to standardize the components of the application environment, including the interface to the operating system, but, in its UNIX developments, it is committed to producing products that meet those standards (see Figure 5.7)

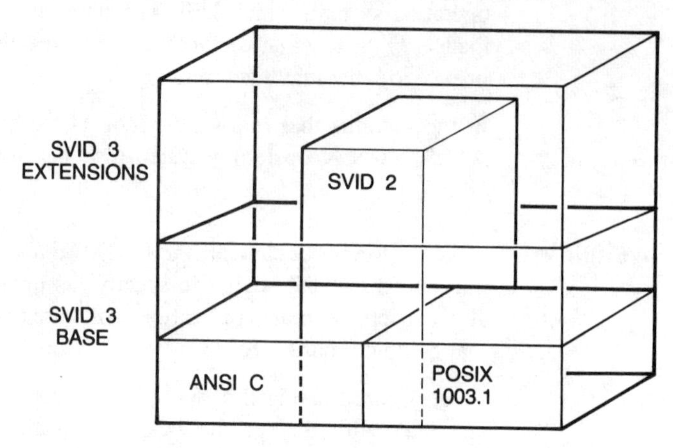

Figure 5.7 Evolution of AT&T's SVID

Since AT&T is also a manufacturer of computer systems and therefore a serious competitor to its own UNIX licensees, and particularly since AT&T took an equity stake in UNIX workstation manufacturer, Sun Microsystems, the potential conflict of interest within AT&T became of increasing concern to many organizations.

Sun commands almost 30 per cent of today's $4.1 billion market for networked workstation and server computers and is both feared and respected by longer-established computer companies. Under the initial agreement between them, AT&T and Sun were to cooperate on the production of future releases of the UNIX operating system. In particular, there was a proposal to move the porting base (the computer architecture used for the source tapes of UNIX when shipped to licensees by AT&T) for future releases of UNIX from AT&T's little-used 32100 microprocessor to Sun's SPARC microprocessor.

Moving the porting base to an architecture which could give a company (Sun) a competitive advantage was unacceptable to many suppliers. The conflict that resulted eventually led those suppliers to form the Open Software Foundation (OSF), on the premise that '*a timely, vendor-neutral decision process*' should result in '*early and equal access to specifications and continuing development*' for the components of the open-system application environment.

Responding to the concerns expressed by its customers, in 1989 AT&T moved the further development of UNIX into a separate division, the *UNIX Software Operation* (*USO*). At the same time, major UNIX licensees, including Unisys, Fujitsu, Xerox, Toshiba and AT&T itself (see Figure 5.8), formed *Unix International* (*UI*), to represent licensees interests on technological specifications, licensing issues and release procedures to USO.

Unisys	Ricoh
Sony	Toshiba
Fujitsu	Dupont
Xerox	Locus Computing
NEC	Relational Technology
Amdahl	NCR
Sun	AT&T

Figure 5.8 (Part) membership of Unix International

The open software foundation

The Open Software Foundation (*OSF*) is an international software development organization founded in 1988 by computer industry sponsors IBM, DEC, Bull, Hewlett Packard, Nixdorf, Apollo, Phillips, Siemens and Hitachi. It was created '*to define specifications, develop leadership software and make available an open, portable application environment*'.

The OSF was incorporated as a non-profit, industry-supported organization in 1988. Membership is open to computer hardware and software suppliers, educational institutions, government agencies, and other interested organizations, worldwide.

The OSF consists of a software development company and a research institute to fund and oversee research that advances OSF's technology. Initial sponsors provided more than $120 m in seed funding and a variety of hitherto proprietary technologies to be used as the basis for further development.

Principles of the OSF include:

- producing software offerings based on international standards;
- using a process which allows early and equal access by members;
- taking decisions in a timely and vendor-neutral manner;
- committing to reasonable and stable licensing terms for its products.

The OSF will base its development on internal research and on technologies selected and licensed from industry participants through a *Request for Technology* procedure. This formal, open process for soliciting input for software developments has so far been used:

- to define the specifications of a graphical user-interface, OSF/MOTIF, expected to become a *de facto* standard;
- for specification of a hardware independent software format Architecture Neutral Distribution Format, ANDF, for mass distribution of applications;
- for specifications of products for inter-working and interoperability requirements.

We look at each of these.

Operating system	POSIX standards	ANSI, ISO, FIPS conformance XPG3, base level
Languages	C	ANSI X3Jl1
	FORTRAN	ANSI X3.9-1978 ISO 1539-1980(E) FIPS 069-1
	PASCAL	ANSI X3J9 ISO 7185-1983 FIPS 109
	Ada	ANSI/MIL 1815A-1985 FIPS 119
	Basic	Minimal Basic ANSI X3.60-1978 FIPS 068-1
		Full Basic ANSI X3.113-1989 FIPS 06802
	COBOL	ANSI X3.23-1985 high level FIPS 021-2
	LISP	Common LISP ANSI X3J13
User interface		OSF/MOTIF
Windowing services		X Window System Version Xl1 ANSI X3113 Libraries: Xlanguage Bindings ANSI X3113
Graphics services		GKS ANSI X3.124-1985 FIPS 120 PHIGS ANSI X3H3.1
Networking services		Selected ARPA/BSD services TCP (MIL-STD-1778), IP (MIL-STD-1777) SMTP (MIL-STD-1781) TELENET (MIL-STD-1782) FTP (MIL-STD-1780) Selected OSI protocols
Database management		SQL ANSI X3.135-1986 (with 10187 Addendum) Levels 1 & 2 FIPS 127

Figure 5.9 OSF's application environment specification (level zero)

OSF/1

While its first products have been targeted at specific requirements for components of the application software environment, the OSF is also developing an operating system, OSF/1. This will be designed to meet POSIX and other international standards, and provide an alternative to AT&T's UNIX. The OFS plans to develop and distribute source code for OSF/1, which will be released in 1990.

An extensive list of base standards, known as Application Environment Specifications, to be used as the basis for OSF/1 development, was published by OSF in 1988 (see Figure 5.9).

OSF/MOTIF

The first Request for Technology (RFT) from the OSF in 1988 was for a windows-oriented graphics-based user interface.

Nearly forty companies submitted proposals. From these, the final selection of the so-called User Environment Component (UEC) of the operating system was made. It consists of three parts: a toolkit based on DEC's X Windows toolkit; a 'look and feel' component from the joint Microsoft/ Hewlett-Packard 3D-windowing system; and a behavioural syntax which follows Microsoft's (and IBM's) Presentation Manager product for the OS/2 operating system.

The OSF development team have integrated these three technologies into a single product, OSF/MOTIF, which has been made available under license to the industry as a whole. The product is sold alone, and will be bundled with OSF/1. It will be portable to other operating systems, under license.

Unix International, which has already recommended AT&T's Open Look as the user interface component of UNIX System V.4, has agreed in addition to support the option of OSF/MOTIF. In return, OSF will separate OSF/MOTIF from its planned operating system OSF/1, in order that licensees may use Open Look with OSF/1 if they so wish.

Architecture Neutral Distribution Format (ANDF)

In 1989, the OSF sent out a Request for Technology (RFT) for the specification and implementation of an Architecture Neutral Distribution Format (ANDF) for open systems. This will enable software products to be distributed on a single magnetic tape format, independent of the machine on which the software is targeted to be run. When and if this is achieved, it could mean that distribution of software would become as simple as the purchase of compact discs or cassettes in the music industry.

Note: In some ways, this could be considered as solving the wrong problem. Many of the problems of incompatibility are caused by the hardware manufacturers' differing treatments of peripherals, such as plotters and printers, and their handling of file input/output. In other words, if the hardware manufacturers could be persuaded to standardize more rigidly first, the problems of defining an ADNF would be minimized.

This approach of the ANDF is opposite to that of the applications binary interface (ABI), reviewed below, where classes of machines using the same processor are designed for total binary compatibility.

Binary compatibility

While formal standards efforts focus on the interface to the operating system and the issues that surround it, a large part of the user community and the companies that service it, would like to see tighter groupings of standards. Their desire is to reduce some of the fragmentation in the industry by providing sets of compatible machines, available from a choice of manufacturers. These can then be sold into the open-systems market in high volume.

The extent of compatibility desired is that of the IBM PC and compatibles, where software written for one machine is guaranteed to run on all others in the same class. This is 'binary compatibility' and, based on the experiences of the PC market, would appear to be a reasonable objective for multi-user machines.

Speaking at Uniforum (an international Unix event held annually in USA) in January 1985, Bill Gates (Chairman and founder of Microsoft Inc.) hypothesized that an installed base of more than 400 000 *binary-compatible* UNIX systems would be required before hardware manufacturers, software authors, and distribution channels could realize the advantages of a large, homogeneous market in the same way as that afforded to the IBM-PC and compatibles market.

By March 1989, Microsoft and its licensees had shipped an estimated 350 000 copies of Xenix for IBM PC and clone architectures. By Gates's criterion then, critical mass has by now been reached.

Binary compatibility has become a crusade for some suppliers in the UNIX marketplace. Like Gates, they believe that the market for UNIX systems will not really take off until it is possible for a computer reseller to stock just one shrink-wrapped copy of a software package, secure in the knowledge that it will run on hardware from almost any manufacturer.

The fragmented UNIX market

Although binary compatibility exists today in the IBM PC and compatibles market, this is far from the case for the UNIX market. For example, a major software company, database supplier Informix Inc., has a product list which runs to 131 pages of distinct products in order to cover 305 types of system as shipped by 72 hardware suppliers.

As Figure 5.10 shows, even computers running UNIX and based on the same microprocessor can, and sometimes do, require different installation media (5.25 in diskettes, 3.5 in diskettes, or cassette tapes), and often present different low-level interfaces to the UNIX operating system. A software developer or reseller wishing to address a significant proportion of the market for UNIX-based hardware must then make heavy investments in recompilation, testing, packaging, inventory and technical support services.

Application binary interfaces

Foremost among the proponents of binary compatibility is AT&T, which in 1987 introduced the concept of the Application Binary Interface (ABI). In an ABI definition, a generic section describes the complete set of services available from an operating system, its associated function libraries, and so on. A

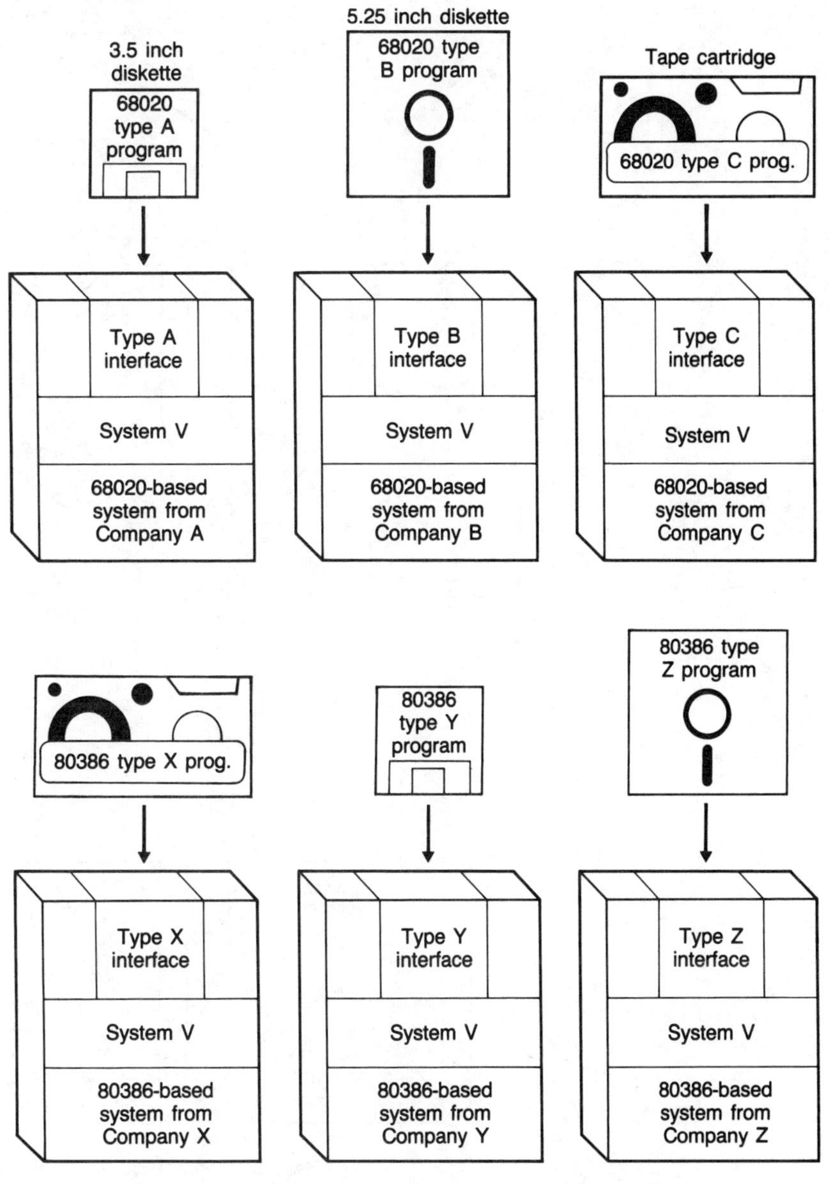

Figure 5.10 The fragmented market for UNIX applications software

processor-specific part gives a precise definition of the manner in which these services are accessed on particular computers. This information, which must be rewritten for each target computer, describes how system functions are activated, how and where parameters should be passed in particular situations, and so on. The information is very detailed: AT&T's draft ABI, which uses Sun's SPARC (Scalable Processor Architecture) as an example, is 271 pages long.

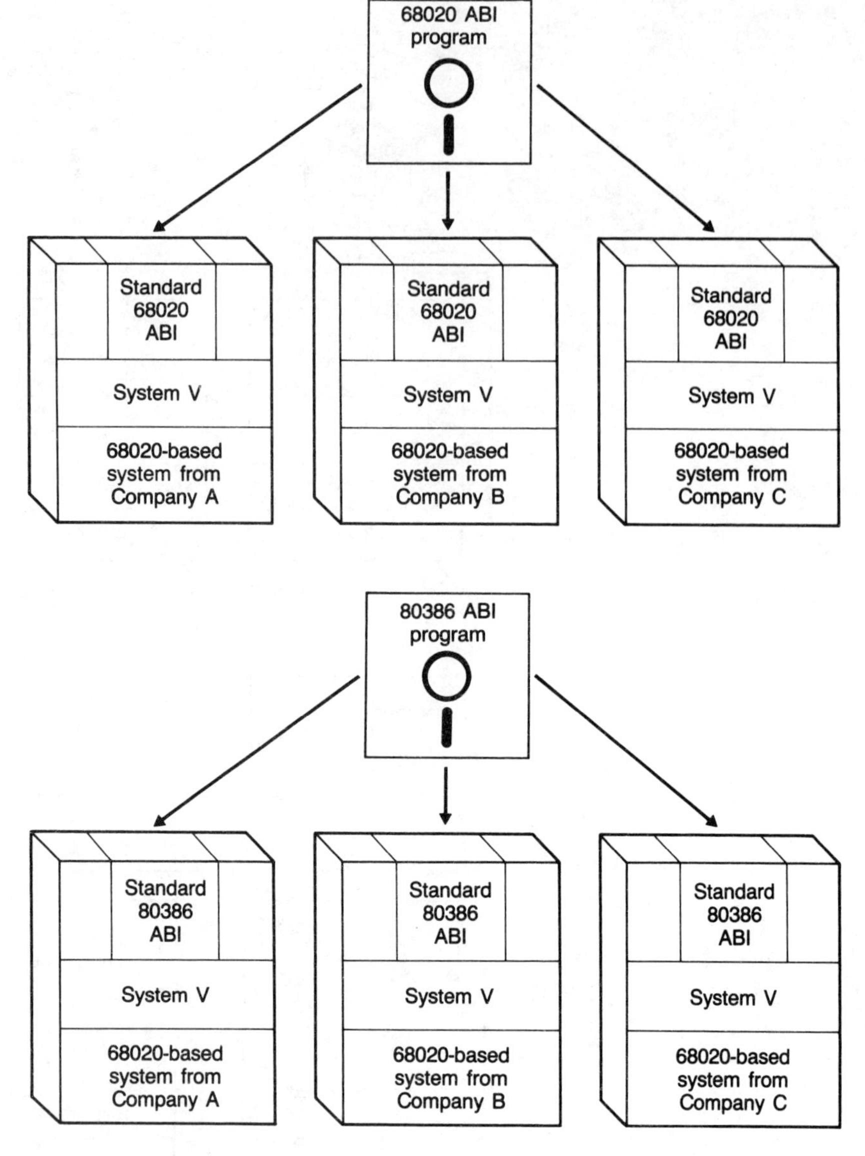

Figure 5.11 *'Shrink wrap compatibility' through ABIs*

AT&T proposed that ABIs be defined for a relatively small number of computer architectures, so producing the situation shown in Figure 5.11, where, for example, a single set of installation media (shown here as diskettes) should load and run an application on *all* Motorola 68020-based systems. A second set should suffice for all Intel 80386-based systems, and so on. This could significantly cut down the number of variants that a software author must produce.

ABIs are currently announced for the Intel 80386, the Intergraph Clipper, Motorola's 68020 and 88100, MIPS' RISC, and Sun's SPARC microprocessor architectures. Although this is a small number of processors, over half the UNIX-based systems so far shipped are covered by it.

Unfortunately, the ABI proposal was seen by some manufacturers as an attempt by AT&T to introduce new technologies, and binary interfaces to those new technologies, ahead of those of its UNIX licensees, thus obtaining an unfair advantage for itself. AT&T added to the concern by announcing ABIs at the same time as it entered its technology agreement with Sun Microsystems.

AT&T made it clear that it would not control ABIs, delegating their production to microprocessor suppliers or their agents. For example, the 88open Group has specified the ABI for the Motorola 88100, while both Intel and MIPS are taking responsibility for ABIs for their own architectures. AT&T has also put mechanisms in place to ensure that all its licensees receive new releases of UNIX source code simultaneously.

Despite this, the OSF has decided not to specify ABIs. Instead, OSF/1 will define application portability only at the source code level. Thus, OSF/1 on its own will not help to ease today's situation, where software compiled, for example, on one Motorola 68020-based system is unlikely to run on another. What is more, application software which complies with (say) the forthcoming 68020 ABI for UNIX System V Release 4, will not necessarily run on a system using the same 68020 processor but which supports OSF/1.

Proprietary architectures

Many computer manufacturers ship their own proprietary processors within their UNIX systems, rather than those supplied by the specialist microprocessor companies. In general, these proprietary architectures are not available for use in other suppliers' products. Examples are HP/Apollo's Prism, HP's Spectrum, IBM's ROMP and DEC's VAX processors.

Since ABIs are intended to bring order to the marketplace when there are multiple suppliers, there is little point in having an ABI for these single-sourced architectures. If a small number of ABIs become well-established, suppliers of systems for which there is no ABI could be shut out of the market, just as personal computer suppliers are shut out if their products are not IBM PC compatible.

It is sometimes argued that the presence of ABIs could stifle innovation, as compatibility becomes more important than the introduction of new technology. This argument was used in the early days of the IBM PC, but the level of innovation and wealth of add-on products and applications for the IBM PC (an ABI market) suggests that these concerns have little foundation. Innovation simply moves to a different level—the application level.

The problem of upward compatibility

In order to preserve the investments that users have made in software, there is a need to continue to support applications running on a particular system *before* that system adopted the ABI. Existing customers are unlikely to upgrade their operating system software, if to do so would stop their existing (non-ABI) applications from working. Unfortunately, ABIs can be difficult to implement for existing processor families.

Because of this, Motorola's 88100 and MIPS' RISC architectures are among the first to have practical ABIs which can be used by software authors. Neither of these new processor families is encumbered by the need to accommodate past practice.

Distribution media

There are two popular media for software distribution for the IBM PC: 5.25 in and 3.5 in floppy disks. A customer ordering a software package for a machine must specify the correct media and a software distributor must stock products on both media. If all else were standardized, this incompatibility alone would cause considerable costs and nuisance value in the distribution channels.

In its original proposal for ABIs, AT&T proposed that for each ABI, a small number of media types and formats (ideally one of each) should be specified for software distribution. A computer which did not support the correct media would not then conform to the ABI. This is a requirement if the goal of 'shrink wrap software distribution' is to be met.

A dealer used to standardization on a single medium in the PC market is reluctant to stock the same binary program for a UNIX application on several different types of diskette and magnetic tape. But the issue of installation media proved so contentious that AT&T retreated from its original position. An ABI now needs only to 'suggest' types and formats. While this means that hardware suppliers do not need to update, replace or supplement today's wide range of incompatible media types, it negates one of the original reasons for the existence of ABIs.

Note: X/Open (see Chapter 7), which, like OSF, has not addressed the issue of ABIs, has had similar difficulties in its attempts to specify a standard medium for the transport of source code between its members' systems.

Software pricing implications

So far, the discussion on ABIs has concentrated on the desire of computer users, software authors, and distribution channels for a less fragmented market, easier and cheaper to service. But there is one important reason for software authors and distribution channels to resist ABIs. It relates to software pricing policies.

The list price for a software package for a particular computer is based broadly on the number of users that the computer can support. For example, a software package for the smallest NCR Tower computer, which supports a maximum of four users, costs much less than the same software package on a 512-user Tower system, despite the fact that the two computers are binary compatible and the two software packages identical. Similarly, a package for a single-user Sun Microsystems workstation is likely to be cheaper than the same package for a multi-user Sun Microsystems computer which runs the same binary code.

There is seldom any technical reason why a package bought for a low-end member of a computer range may not be run on the high-end machine. Only the legal constraints of the licensing agreement bind the user.

In UNIX System V Release 4, the UNIX operating system provides a mechanism (named streams) which could monitor, and limit, the number of concurrent users of a software application. Current versions of UNIX make such control difficult, and so few software vendors have attempted to implement them. Instead, they expect their customers to comply honestly with their pricing rules.

A widespread move to ABIs will weaken the ability of the software authors, and of their distribution channels, to enforce this differential pricing. Technical barriers to the number of users served by a given package will be severely reduced. This could reduce the UNIX software market to the state of the PC software market where, having bought a package for a fixed (usually low) price, a user may choose to run it on a slow, old PC-XT, or on a new Intel 80386-based system with many times the power.

PC software suppliers are attempting to charge in various ways for the use of software by multiple users. For example, methods have been developed to control the number of connections made to networked versions of their software. To date, most methods are unsatisfactory, both to the suppliers and to the users.

Microsoft's policy for Xenix

While there are several factors working against the rapid adoption of ABIs by the UNIX hardware and software suppliers, Microsoft's Xenix/UNIX for Intel's 80386 microprocessor is the first UNIX environment to reach 400 000 binary-compatible units shipped to date.

This is not because the 80386 version of UNIX has shipped that many units, but because Microsoft's policy has always been to preserve strict binary compatibility between one release of Xenix and the next (see Figure 5.12). Thus software developed for one version of Xenix was guaranteed to run on the next, albeit without using any of the new features.

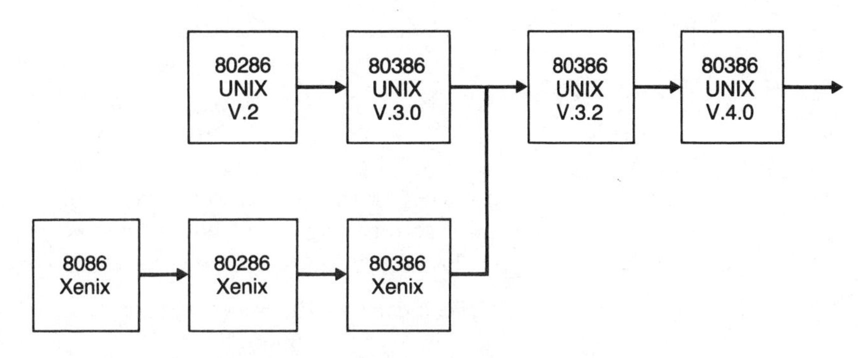

Figure 5.12 Binary compatibility for Xenix

Following an agreement between Microsoft and AT&T in 1987, this compatibility was carried forward into the current release of UNIX System V, and will be preserved in the future as UNIX continues to evolve. This means that UNIX System V for the Intel 80386 can run binary application programs from *all* previous versions of Xenix for Intel 8086 family processors, and from previous releases of System V for the 80286 and 80386.

When UNIX System V sales on Intel processors by other suppliers such as AT&T, Bell Technologies, Interactive Systems, Microport and others are added to Microsoft (and its licensee, SCO) Xenix sales, critical mass for the PC UNIX market has almost certainly been attained. All predictions show that growth will continue to be rapid, and those companies who have *not* adopted an ABI must be watching it very closely.

Open DeskTop from SCO

The last few years have witnessed a rapid increase in the power available from desktop and personal computers. As the performance of these systems grows, users want to harness this new power, and applications are growing ever more complex.

The complexity and functionality required has been pushing operating systems to the limit. Software developers, frustrated by the limitations of Microsoft's MS-DOS and its delay in producing OS/2 for the Intel 386 and 486 microprocessors, have increasingly turned to the UNIX operating system (see Figure 5.13).

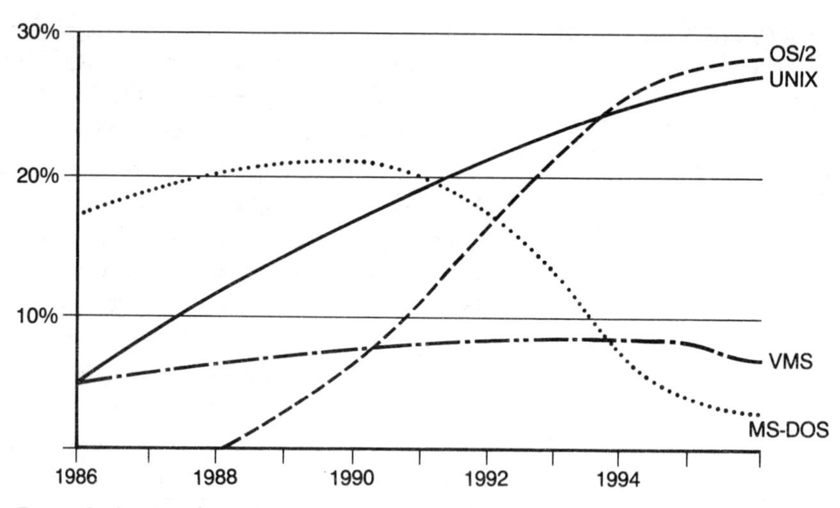

Figure 5.13 Trends in operating systems

To take advantage of this, the Santa Cruz Operation (SCO), under license to Microsoft, packaged up UNIX for the PC market, and is largely responsible for the large installed base of PCs running UNIX.

Concerned about the need for binary compatibility for the more sophisticated machines of the future, SCO has cooperated with several companies to produce a single product for the application environment for machines using

DOS compatibility with:

Merge 386 DOS Development System (PCiLIB) from Locus

Database management with:

Ingres/386 SQL Code Generation System
Preprocessor for C
Ingres GCA Library
SCO ISAM Libraries

Many of the development tools and APIs are based on industry standards and the following are provided with the development system:

- for UNIX: CodeView, ADB and SDB debuggers; Microsoft and AT&T C language standard; UNIX System libc and libm libraries;
- for Xsight: OSF/Motif API (presentation description language and toolkit);
- for Ingres/386 SQL: SQL and GCA libraries;
- for TCP/IP: The socket library and TLI (and TLI for streams development);
- for LAN Manager: NetBIOS/Int 5C programming library;
- for DOS and OS/2: cross-development libraries;

SCO UNIX System V/386, release 3.2 with:

X/Open XVS Libraries
IEEE, POSIX and FIPS POSIX Libraries
BSD Select, Sockets, Reliable Signals
CodeView C Debugger
MASM Assembler
DOS and OS/2 Cross Development Tools

Networking with:

TCP/IP Development System
NFS Development System

X Window System with:

Xsight X Window System Version 11 Development System (based on XUI from DEC)

OSF/Motif Software Development Package with:

X libraries, X server and clients
OSF/Motif toolkit
OSF/Motif window manager
OSF/Motif style guide (Presentation Manager)
OSF/ Window Widgets
Presentation Description Language (PDL)

Desktop Manager:

To match X.desktop from IXI

- Options to replace Ingres by either Informix or Oracle are provided.

Figure 5.14 The components of Open DeskTop

the Intel microprocessors. Based on current *de facto* standards, this product, known as Open DeskTop (ODT) will be supplied by a number of major manufacturers as the application environment for their low-end Intel-based UNIX machines.

Open DeskTop is a collection of development tools and utilities integrated with the UNIX operating system, and provides a full windowing UNIX system for an Intel 80386-based PC. It is supplied by SCO as a complete software suite and includes a standard graphical interface based on OSF/Motif. It also provides SQL distributed database, communications and networking facilities. Open DeskTop therefore has a range of facilities directly comparable with those of Microsoft's OS/2 Extended Edition running Presentation Manager, designed for single-user systems.

The basic difference between Open DeskTop and OS/2 lies in the fact that, whereas OS/2 is a proprietary Microsoft product, ODT is built of openly available components that conform to international standards (see Figure 5.14). As a result, according to SCO, ODT has taken approximately one year from inception to delivery, compared with three taken so far for OS/2. This clearly demonstrates the efficiency and cost effectiveness that can be obtained by using compatible (standard based) products.

It is interesting to note that Microsoft took an equity investment in SCO around the time of the ODT announcement, thus confirming to many people the validity of SCO's approach.

The X/Open Common Applications Environment

The X/Open Company (see Chapter 7) was formed in 1984 to define a common applications environment (CAE), comprising all the elements of a computer system that need to be standardized in order to achieve the objectives of open systems.

The CAE is made up of standard components, where such exist, and from the best that the computer industry has to offer otherwise. The CAE is thus a moving entity, constantly expanding and changing as technology advances.

Members of X/Open, which includes twenty of the world's largest computer manufacturers, are committed to produce systems that support the CAE. Many of the international software companies have developed products to run on such systems, and increasingly, users are specifying the CAE in their procurement guidelines. In principle, though not yet in practice, the CAE is the most important standard that exists today.

SCO's Open DeskTop is a subset of X/Open's CAE, being conformant to it, but being specific both to the Intel processor, and to particular products in its implementation.

Summary: standards for portability

In this chapter, we have examined the standards that exist to aid the implementation of portability. These consist of standards for the operating system and related areas that make up the total environment in which the apl tions software must run.

There is a set of internationally supported standards for the applications environment emerging under the POSIX banner, but these barely begin to tackle the totality of standards needed. Many related standards are required in areas such as security, data handling, networking and internationalization, as well as in software distribution media and peripheral support.

While a *de facto* standard for the operating system which implements the internationally approved POSIX standard exists in the form of the UNIX operating system from AT&T, a competitive product is planned from the Open Software Foundation, and others may emerge from the proprietary products in existence.

While this may be attractive to the hardware manufacturers, particularly those which are members of the Open Software Foundation, it is unlikely that having several competing products for the standard operating system in the open systems market will benefit the user.

If manufacturers continue to add their own variations to otherwise standard products, the open-systems market will remain fragmented, to no one's ultimate advantage. If these are indeed useful, they should be incorporated into the agreed standards.

The emergence of a single, dominant grouping of standard components on subsets of machines using the same microprocessor and which satisfy both the market's technical requirements and commercial interests could lead to a *de facto* standard being adopted for the complete application environment.

The offering of Open DeskTop from SCO brings together a number of open systems standard products (both *de facto* and *de jure*) for systems using the Intel microprocessors, and may well influence the market significantly in the future.

The most important standard, in principle if not yet in fact, is the Common Applications Environment from X/Open. When manufacturers are shipping systems in volume that fully support X/Open's CAE, users may finally get what they both want and need.

Standards for interconnection

In the 1970s, computer manufacturers developed and marketed proprietary networking products to allow communications between their own computers. Although many of these were technologically innovative, they suffered from the fundamental flaw that they were generally only appropriate for connecting together machines from a single manufacturer.

As users move to multi-vendor purchasing strategies, it is increasingly important that a standard communications technology be available to interconnect machines, independent of the manufacturer. Networks can then be built with machines from a variety of suppliers.

The ISO standard for open-systems interconnection (OSI) is such a standard, and underlies all the developments in networking of the last decade. But while the OSI work may have unified thinking on open-systems interconnection, it has not simplified the issues. On user sites, OSI products must learn to co-exist with highly-developed, *de facto* standard communications products, such as IBM's SNA and the US Department of Defense TCP/IP protocols. And OSI itself has given rise to a bewildering variety of international standards, with many more in the pipeline.

In the late 1990s, companies need to realize the economic benefits that will come from integrating systems both within their own organizations and across systems within their suppliers' and customers' organizations. Efficient integration requires agreement on standards for applications such as electronic mail, document definition, file management, and the general area of electronic data interchange (EDI). Some of these requirements for application standards are common to all industries and some are specific to particular vertical segments.

All of the traffic carried on communications networks—voice, telex, facsimile, even video—can be translated into digital data for transmission. As a result the 1990s will see the widespread introduction by telephone companies of integrated service digital network (ISDN) connections. These will replace the older voice-only circuits and provide consumers with digital connections capable of carrying any type of information.

In this chapter, we will review standards for interconnection, examining the internationally approved OSI standards in some detail, and relating these as much as possible to the important *de facto* standards. We will extend the discussion to important application standards, especially those related to EDI, and conclude with mention of some important groups of standards which are required for particular vertical applications.

Open-systems interconnection

When a group of people speak various languages but have a need to communicate, as in the United Nations, for example, it is impossible for everyone to learn the language of everyone else. A standard language, usually English, is therefore agreed. People who can speak that standard language speak it directly, and those who cannot, speak through interpreters, and can listen to a translation of the proceedings into their own language.

If the translation is done smoothly, each person will hear (over the earphones) a debate in his or her own language, and will be able to make contributions in that language.

Computers and IT systems are often in this situation. In order to deal with it, different manufacturers have developed various means of communication, usually specifically for their own computers. However, because companies now need their computers to communicate universally, the analog of the standard language, the techniques of interpretation and translation need to be developed for computers and adopted by the manufacturers.

In concept, the solution for computers is very similar to that for people. A common 'language' is agreed, which allows computer systems to send intelligible information to each other, despite being built in different ways. This situation is shown in Figure 6.1.

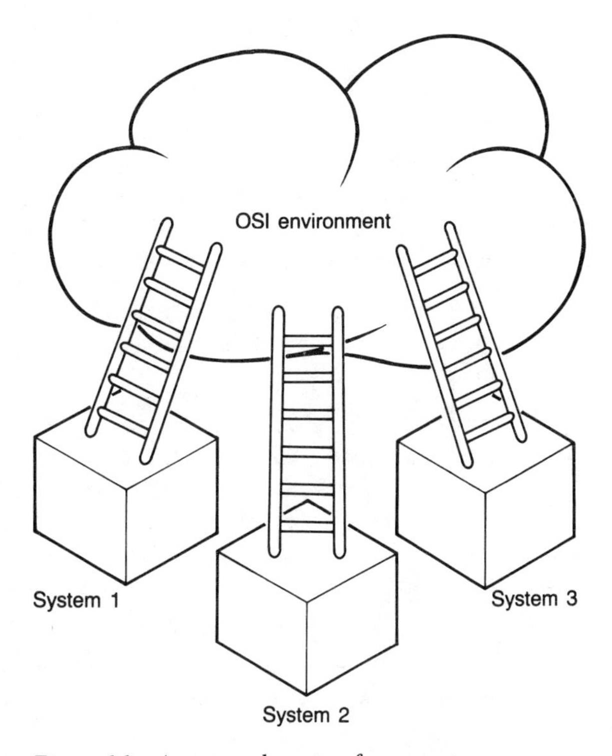

Figure 6.1 A common language for computers

In Figure 6.1, System 1 communicates with System 2 by sending a message which is first translated into the standard format within System 1, and then sent over the communications network to System 2. There, the message is translated back from the standard format into a format that System 2 can understand.

As with the human translation system, an individual system is free to conduct its own internal business in any language it chooses, as long as it maintains a highly standard interface to the common environment.

Public standards for interconnection

In principal, if agreement could have been reached on a *de facto* standard and that standard adopted by the computer industry it would have been possible for a communications standard to have been created from a proprietary product in the marketplace. This would have required the manufacturer of the product to be willing to license it to the rest of the manufacturers, of course.

However, as we saw from examples in the previous chapter, the potential competitive advantage that could be gained by such a 'standard maker' is considered to be so high that the industry finds it difficult to reach consensus, and usually prefers for there to be several competing products with the potential to conform to the standards.

In the area of communications, the result of the competitive forces was such that, although there were potential candidates for a public standard from several manufacturers' products, little progress on an international standard was made until the public standards-making bodies became directly involved. ISO began to coordinate the work, which resulted in the international standard known as the OSI seven-layer model, in 1977.

The OSI seven-layer model

The process of communication between two systems can be visualized as made up of several separate tasks, some carried out by the 'speaker' and some by the 'listener'. These tasks can be thought of as separate modules, operating in hierarchies within the two system. The modules can then be regarded as a series of layers on each system, through which messages must pass.

In the ISO approach, the communications task was divided up into seven layers. For this reason, the resulting OSI scheme is known as the 'seven-layer model'.

The two main layers

As Figure 6.2 shows, in the OSI seven-layer model, communications breaks into two major tasks. The first of these is carried out by the lower three layers, under the heading of 'interconnection', or 'networking'. The second, carried out by the top three layers, is referred to as 'interworking' or 'interoperation'. The fourth layer (the transport layer) acts as an interface between the two tasks.

As we shall see, in the lower half of the seven-layer model (layers 1–3), ISO standards for both local and wide area networking are well developed and widely adopted. Although local and wide area networks behave differently, the transport layer can hide the differences. The top layers can then use either type of network for communications.

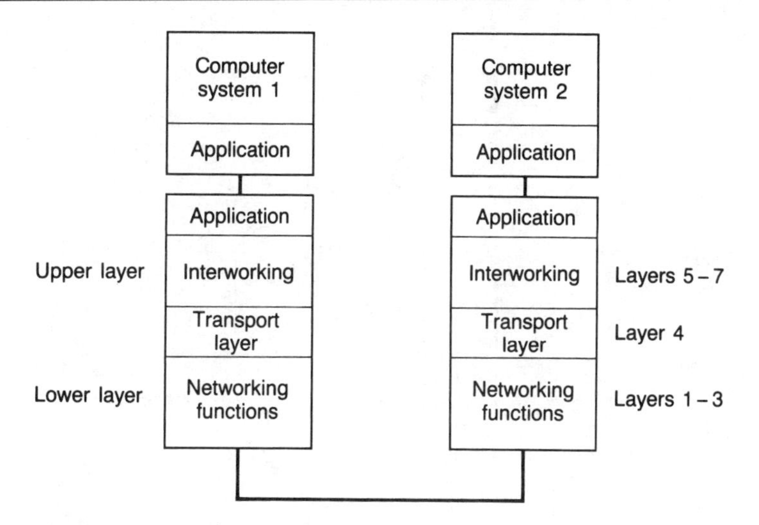

Figure 6.2 The two major OSI layers

The transport layer defines a standard set of services which must be provided by systems manufacturers, and which are then available for applications software developers. This modular approach makes it possible to build transport layers for non-standard networks so that they may use standard upper layers.

Because the transport layer provides the division between the two major tasks of communications and applications, companies providing products for the implementation of OSI standards often focus on it. For example, AT&T includes a transport layer interface (TLI) in the network services extension to its UNIX System V Interface Definition. This was extended and refined by X/Open in its X/Open transport interface (XTI).

Both TLI and XTI are interpretations of the OSI standard. They are instructions to system builders on how to provide a correct transport service, and to software developers on how to use that service correctly in the applications they build. Microsoft's OS/2 operating system includes a similar transport service specification.

The upper half (levels 5–7) of the seven layer model is more difficult to standardize. Different upper layer standards have to be developed for applications such as electronic mail, file transfer and document interchange. These standards are only now beginning to mature and to be adopted by manufacturers and users.

Figure 6.3 illustrates the general situation for implementation and application of the two main layers in the model, including the technologies available for interconnection, and the applications that are of common interest. We will discuss these in more detail later.

Options in the OSI standards

In order to accommodate the wide variety of applications needed, there are many options within the OSI standards. Different systems may select different sets of options from within the OSI standard, giving rise to the possibility

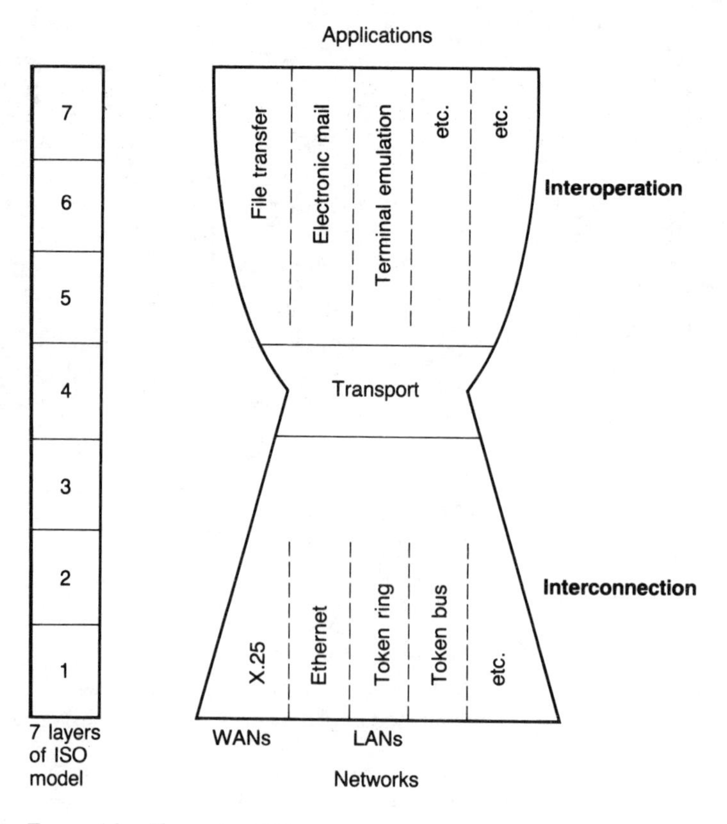

Figure 6.3 The main OSI layers

of incompatibilities between implementations. Much standards activity centres on keeping these standards compatible as they evolve—a process known as *harmonization.*

Both users and systems builders can easily become confused by the multiplicity of options within the OSI model. There is therefore a need to direct the user to those parts of the standards which should be adopted in particular circumstances.

To deal with this, AT&T's XTI specifies some sections of the transport service as mandatory and others as optional, while X/Open advises application developers to build applications which either use only the mandatory functions, or else can adapt their behaviour according to the richer sets of functions provided by particular systems.

Detail of the seven-layer model

When two computer systems are using OSI standards, each may have its own operating system, application software, and OSI interface. To start the process of communication, the application software from one system must send instructions to its OSI interface, requesting a communications service.

Messages go to the top of the seven layers (see Figure 6.4), and are passed successively down through the layers to the lowest level. This sends them across the communications medium to the other system, where they are passed up through the seven layers to the second system's application software. In the process, the seven layers each perform their own function.

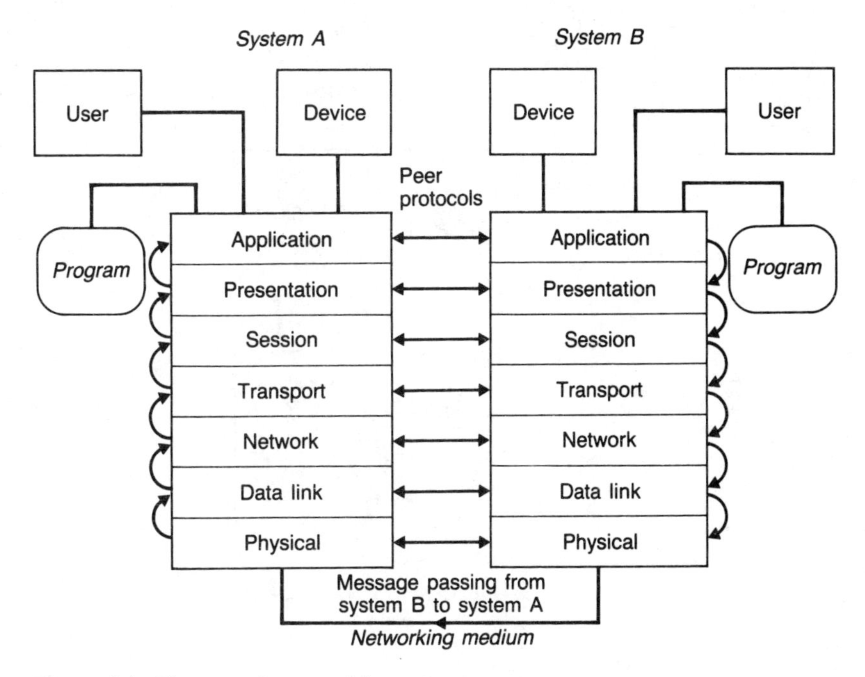

Figure 6.4 The seven-layer model

Each layer performs a service for the one above by requesting services from the one below (see Figure 6.5). Communication between systems is made by peer-to-peer communication within comparable OSI layers in the two systems. These communications with peer layers are effected by tags, which are added to the data in one system, then removed by the equivalent layer on the other system. These tags work a little like envelopes on letters. The letter is enclosed by an envelope and passed to the next lower layer during dispatch. Following receipt the envelope is removed before the message can be acted on by the next higher layer.

For each layer in the seven-layer model, the ISO standard defines what the layer must be able to do, and specifies the way in which this should be done.

The lowest layer, *the physical layer*, is directly connected to the physical medium (wires usually) between the systems. This layer sends and receives a stream of bits across the medium, providing a physical connection between the two systems. It also specifies the electrical characteristics of the interface.

The second layer, *the data link layer*, controls the flow of data, and the correction and detection of errors. In local area networks, this includes control of the access to the medium, i.e. deciding which system can use the network.

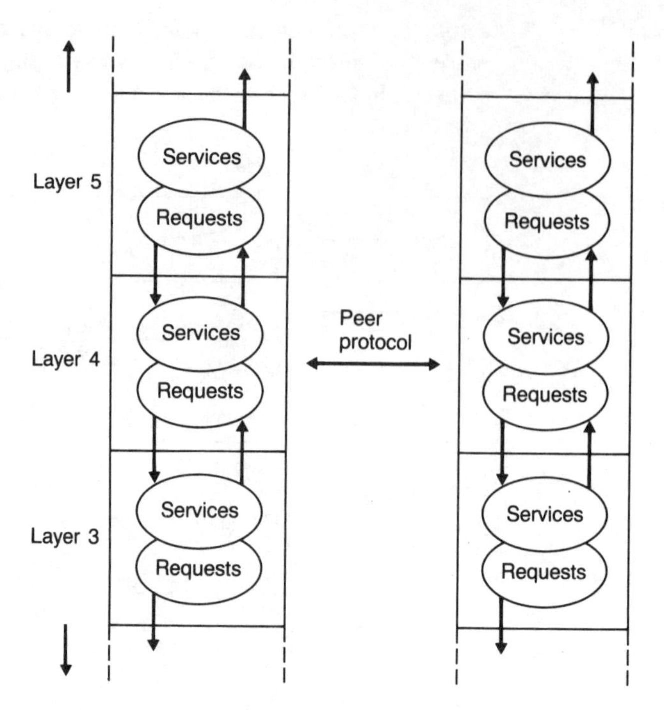

Figure 6.5 Communications between OSI layers

The third layer, *the network layer*, controls routing and other functions, including flow control and sequencing. If there is no direct connection between the two systems which wish to communicate, the network layer finds out what intermediate systems can relay the messages to their destination or, if necessary, what other networks there are which can carry the connection part of the way.

Where two networks differ, they can be connected by an intermediate 'gateway' system, which has the lower three layers of each network. Communications then pass up through one half of the layers and down through the other (see Figure 6.6)

The fourth layer, *the transport layer*, is the join between the upper and lower parts of the model. It provides and monitors the quality of service and error rate, so that specialized functions in the upper layers can rely on a flow of data. Some transport layer options allow several messages to be multiplexed—that is, sent down one network connection at the same time, in different time slots. This is also a function of X.25, i.e. layer 3, discussed later.

The fifth layer, *the session layer*, arranges 'conversation' between the systems. Session layers on the end-systems agree the way data should be sent—in both directions at once, in alternate directions, or in one direction, for as long as the connection is required.

The sixth layer, *the presentation layer*, is concerned with character sets and symbols. The presentation layers of the two systems agree on a common

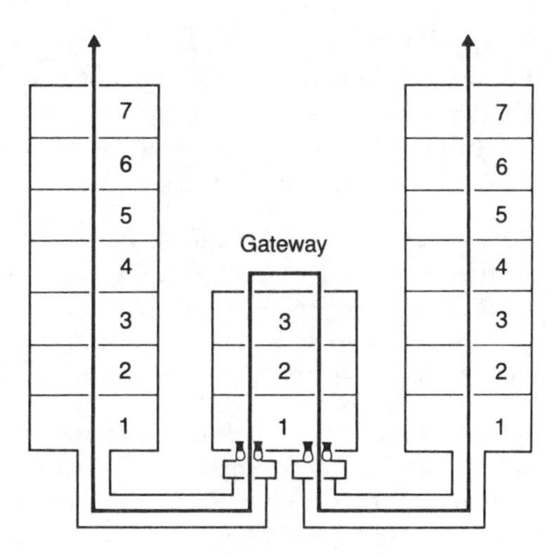

Figure 6.6 Gateways for OSI systems

representation—basically an alphabet and language for the higher software to communicate in.

The seventh layer, *the application layer*, handles semantics—the meaning of the communication. It is specialized for different applications, but is not an application in itself. For instance, a seventh layer product for electronic mail is not in itself an electronic-mail system. It provides the central services that an electronic-mail system will need. User interfaces and other supporting codes are required for an actual software application.

Standards for the OSI layers

Working in cooperation with the CCITT, IEEE, and many other bodies around the world, ISO has produced and is producing almost continually, standards for the various OSI layers. Some of these have been (or are being) widely adopted.

At the lower levels, ISO has adopted local area network (LAN) standards developed by the IEEE. These include Ethernet and Token Ring technologies, and are known after the committee which defined them as the *802 standards*.

The wide area network (WAN) standards were taken from the CCITT and are based on packet switching technology, carried over a variety of line types, and typified by the CCITT's X.25 standard. The main differences between these two networking technologies are the kinds of cables used, and the means by which access to the network is controlled.

Ethernet and X.25 are discussed in more detail later.

Standards for the OSI transport layer define five levels of service, in Classes 0–4:

- Class 0 is a basic service with few features;
- Class 1 includes recovery after the link is lost;

- Class 2 includes multiplexing;
- Class 3 includes recovery and multiplexing;
- Class 4 is specially tailored to operate over so-called 'connectionless' networks (those having the ability to exchange network messages without having a specific data link connection).

There are published ISO standards for the session and presentation layers. In addition, there is a related standard, the Abstract Syntax Notation (ASN) which is designed to help presentation and application layers to agree a syntax in which to communicate. ASN is also a language in its own right, in which higher layer standards can be specified.

At the application layer, a different standard is needed for each main application task required by the applications software. The most important standards so far agreed are X.400 for electronic mail, FTAM for file transfer access and management, and more recently, VT for virtual terminal protocols. Electronic mail, file transfer and terminal emulation are basic functions which can be used by developers as a base on which to build larger applications.

Other important standards close to finalization are office document architecture (ODA), and job transfer and manipulation (JTM).

Before discussing the OSI application level standards, we will overview the *de facto* standards that exist for networking technologies in order to show how they fit into the international OSI standards work.

De facto standards for interconnection

There are two main technologies in widespread use today (1990) for Interconnection. One, the TCP/IP technology, originated from the US Department of Defense, and is a semi-public standard in that its specification is in the public domain. Products supporting it have been developed by many manufacturers, and these are in widespread use in many multi-vendor networks. The other, IBM's SNA technology, is a proprietary technology, owned and controlled by IBM, and widely used in the sense that almost any product from market leader IBM is widely used.

Transport Control Protocol/Internet Protocol (TCP/IP)

Transport Control Protocol/Internet Protocol (TCP/IP) is a widely used *de facto* standard for communications between computers. It consists of two parts: the Transport Control Protocol, which handles functions approximately the same as those of Layer 4 (transport) of OSI, and the Internet Protocol, which provides services approximately equivalent to those specified for OSI Layer 3 (network). However, TCP/IP is not a standard implementation of those OSI layers.

Although TCP/IP was developed for the US Department of Defense (DoD), the DoD now insists on OSI specifications in new contracts.

Currently, TCP/IP is provided in many UNIX systems, especially those requiring to communicate over local area networks. It is also used in some

PC networks. In the short to medium term it will continue to be used, mainly because it has been integrated with so many applications.

Ultimately, TCP/IP is expected to be replaced by the OSI transport and network layers. As evidence of this, OSI communications have now been incorporated into AT&T's UNIX products, the OSF's basic set of standards and X/Open's Common Application Environment.

IBM's SNA

Systems Network Architecture (SNA) is a set of IBM proprietary communications products which at one time was regarded as potentially competitive with OSI. Having a similar layered architecture, it was designed by IBM to allow its own disparate computer architectures to communicate with each other.

SNA could not become an open communications standard for the computer industry because IBM was not willing to give control of its specification and further development to a public standards body. Instead, in response to user

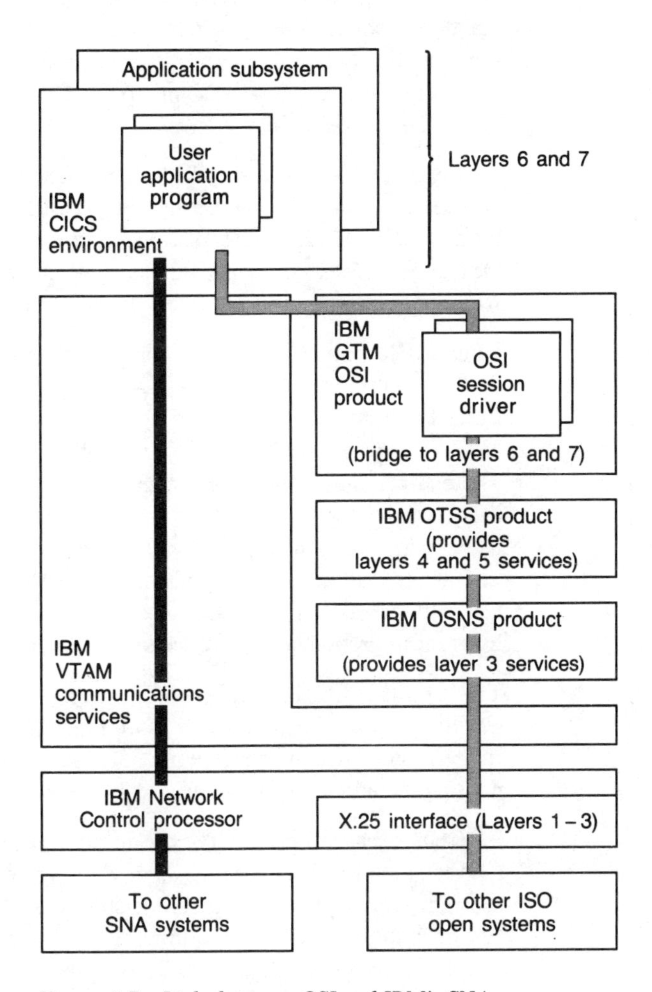

Figure 6.7 Links between OSI and IBM's SNA

demand for open standards, IBM has produced connections between SNA and OSI installations, and provides products which allow OSI communications over its SNA links. Figure 6.7 illustrates the current situation.

Within its own product line, IBM is committed to provide common communications, user interface and programming interface across its entire computer range, in an overall, proprietary architecture it calls System Application Architecture (SAA). Originally, the communications part of SAA was to be SNA, but OSI is now also included.

Many existing IBM users who have invested heavily in SNA now wish to migrate to the international OSI standards. This will be increasingly necessary if they are to communicate with the rest of the (non-IBM) world. Moving a large installation from SNA to OSI is likely to be expensive in the short term, though may save costs in the longer term. This is the case with most open-system migration strategies.

Gateways between SNA and other communications environments do exist, but direct connections generally provide better performance and increased functionality For this reason, many users require both SNA and OSI protocols, sometimes even in the same machine. Providing this may be difficult for some computer suppliers.

Although the lower layers of its proprietary LAN, the Token Ring, have become integrated into OSI standards, IBM has not yet indicated support for the higher level OSI protocols over its own LANs. By providing wide-area OSI only, IBM is allowing users to link externally with non-SNA systems, while encouraging them to use SNA within their own sites.

Unless IBM policy in this area changes, the problem of complete integration of IBM proprietary networking technologies into OSI standard networks is likely to remain.

Other networking technologies

Local area networks

There are many networking users today not yet using OSI technologies. Many of these are users of *de facto* standards, such as Ethernet and Token Ring, but are not using the higher layer OSI protocols. Some may plan to migrate to OSI later, while others have installed networks as part of a proprietary network architecture. Because of their widespread use, and the implications for migration strategies, we look at them briefly.

Ethernet networks

The original Ethernet technology, which was developed by the Xerox Corporation, used coaxial cable, and transmitted data at 10 Mbits per second. In the original design, an Ethernet network could support a maximum of 1024 nodes, on a number of sections of cable, each up to 500 m long.

Ethernet is based on carrier sense multiple access/ collision detect (CSMA/CD) technology. Here, each system wanting access to the network checks to see if it is available. If it is, the system may transmit. However, if another system transmits at the same time, the two messages will collide. If a collision is detected, both systems allow a randomly generated time interval, and try again.

Variants to the original design have been developed and subsequently approved as standards by the IEEE. These can run over a variety of media, including twisted pair wiring (as in telephone) and fibre optic cables. Usually the cheaper the media the shorter the range, or the slower the transmission speed.

Ethernet networks were designed for office and general purpose use. They are now provided by more than 200 suppliers worldwide. Although widely used, market share is beginning to slip as that of the more recent arrival, Token Ring, expands. Over time, the two standards are predicted to reach approximately the same percentage market share.

Token Ring

Token Ring was designed by IBM as a general purpose LAN, transmitting at 4 Mbit/s over twisted pair wiring. IBM now supplies a 16 Mbit/s version.

With Token Ring, systems are on a ring network and pass a token from one to another continually. Only when a system is in possession of the token is it allowed to transmit on to the network.

A number of other manufacturers have adopted IBM Token Ring technology or offer products to connect IBM Token Ring networks into their own proprietary networks. As a consequence, the use of Token Ring has grown fast, and it is now considered a *de facto* standard.

Token Bus

Token Bus is a networking technology designed for harsh environments where a guaranteed response time is required, and is therefore specified for manufacturing applications in the OSI standards. It transmits at 10 Mbit/s on broadband (a signalling technique which sends multiple radio frequency signals along a cable) cables, and again uses a token system to control access to the network. As with Ethernet, cheaper alternative cabling can be used if desired.

Fibre Distributed Data Interface

An important standard for the networking in the future is predicted to be the fibre distributed data interface (FDDI). This standard is being prepared by ANSI. It will use optical-fibre cabling, and is expected to be used as a backbone for LANS when today's Ethernet and Token Ring systems become overloaded. Since FDDI can carry both Ethernet and Token Ring traffic within its 100 Mbit/s bandwidth, it should be able to handle the demands put on the network by future distributed and graphics-based applications.

Light signals over an optical fibre have a longer range than electrical signals over a cable. As a result, FDDI networks can extend as far as 100 km, thus making the concept of 'metropolitan area networks' possible for the future.

The FDDI standard is proving difficult to define, and products conforming to the final version of it are not expected to be available until mid-1990 at the earliest. As usual, there is a danger that early implementors may be locked into a non-standard version,in which case upgrading to the full standard later may then be difficult.

Wide area networks

Wide area networks (WANs) cover long distances, often worldwide. Messages are carried on a variety of media, including cables, telephone lines, public data networks, and radio signals. The lower levels of the networks include devices, such as modems, which handle the basic signal transmission over the different media.

X.25 is a CCITT recommendation for packet switching over a wide area network. It defines the network up to layer 3 of the OSI seven-layer model. It specifies the protocol used by 'packet terminals' connecting to a packet network, i.e. the connection between the terminal or host itself and the packet node or switch. Protocols used for connection between the many nodes on a packet switched network are often proprietary, although some do use the CCITT's X.75 specification (which was originally devised for international gateways).

Packet switched networks carry data in units with a fixed maximum size, called packets. Using packets prevents a particular message monopolizing part of the network—large messages are broken up into packets and then reconstructed at their destination. When several messages are sent at once, the packets are alternated. The fact that information is sent in this disjointed fashion is hidden from the system, which sees a continuous connection.

Packets are assembled and put on to the network, either by the end-system itself, or by a device called a PAD (packet assembler/disassembler). How these packets are routed across a network is shown in Figure 6.8.

PADs are used to *assemble* data into packets which are submitted to the network, and to *disassemble* incoming messages. PADs also 'buffer' messages, storing them as necessary to cope with the differences between the transmission speed of the network and that of the system. PADs can usually handle several systems at once, multiplexing them into one network node. PAD functions are defined by CCITT triple-X standards, X3, X28 and X29.

OSI: the application layers

We saw earlier that the application layer of OSI is the interface between the OSI software and the applications software. Standards for it can never be complete, because new application requirements are constantly emerging.

However, there are a number of standards that have so far been developed for special purposes and are available in products from a number of suppliers. The most important of these are discussed below.

File transfer access and management, FTAM

Computers use organized chunks of data, called files, which are usually stored on devices such as disks on a system. These files are kept in an orderly way by the system. When they are moved around or altered within the system, this is referred to as 'file-handling'.

The way in which files are handled varies between systems from different manufacturers. This can cause problems when an organization needs to transfer information reliably from one computer to another, or if a user wishes to access from one machine, information stored on another.

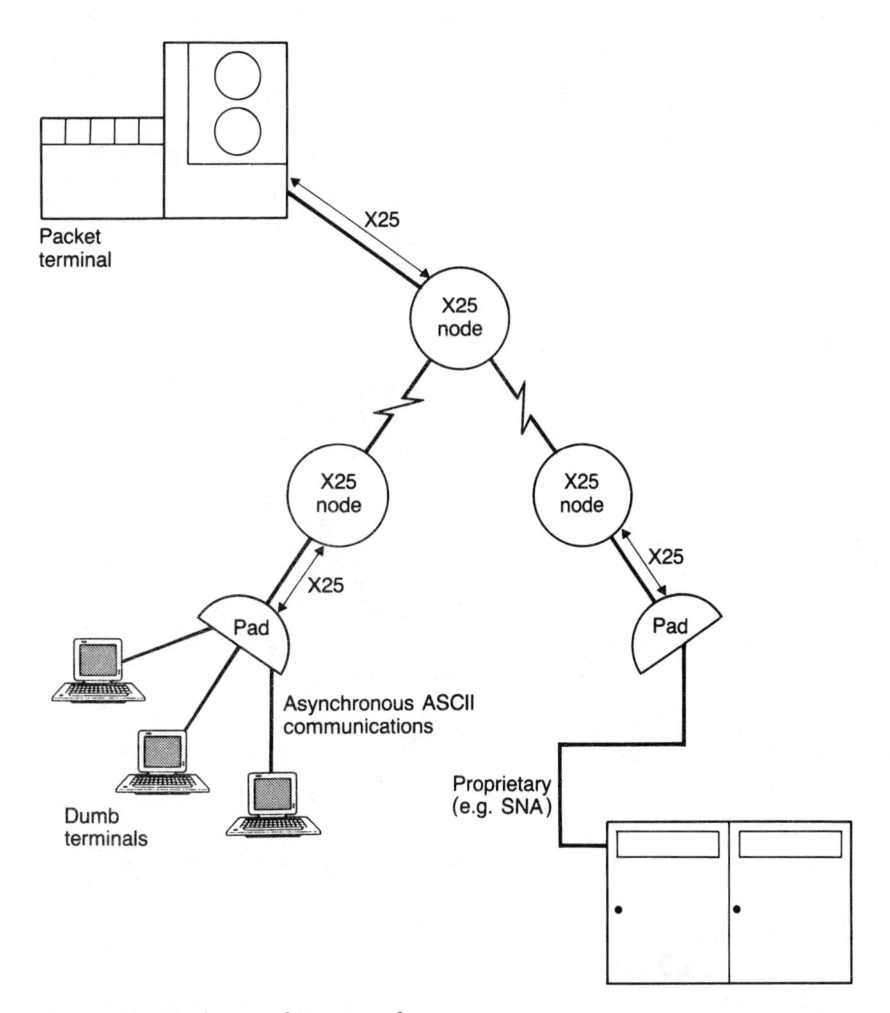

Figure 6.8 Packet switching network

The file transfer access and management (FTAM) standard (see Figure 6.9) solves this problem, by allowing transfer of data-files, or parts of files, from one system to another, even if they have totally different file structures and data-handling methods. It also allows users to carry out file management on remote systems.

The FTAM standard guarantees that either a correct copy of the file or a message indicating failure is sent. If there are problems, the transfer restarts automatically, until a correct copy is received. FTAM can be used by people, who may need access to a file on a remote machine, or by programs, which may transfer large numbers of files at a time.

Some current FTAM implementations do not implement the full ISO standard. This is sometimes because of limitations in the host machine. Users need to check the features that are provided to determine whether it is

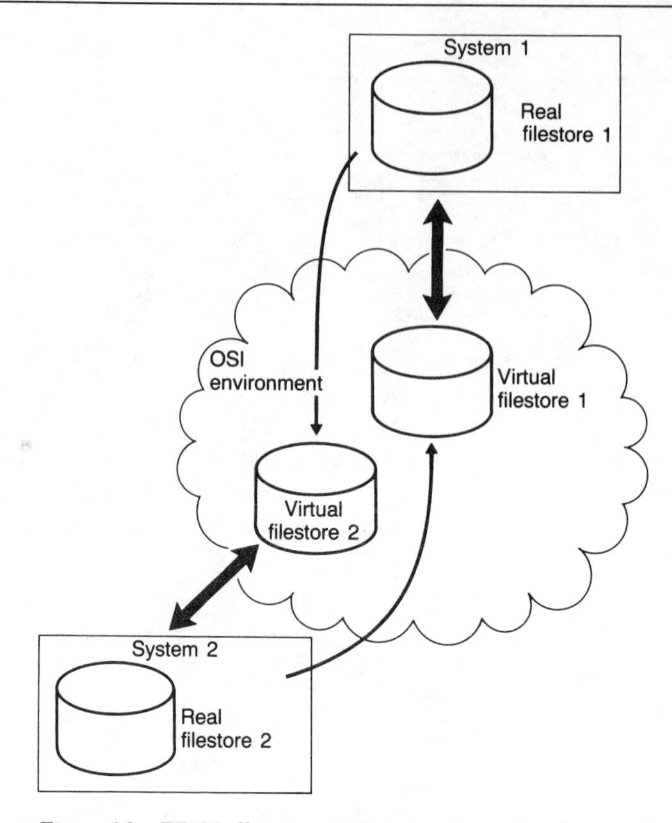

Figure 6.9 FTAM allows manipulation and transfer of remote files

possible to upgrade to the full standard, because it is possible that the subset of FTAM supplied will restrict communications with some systems.

Message handling systems: X.400 electronic mail

In today's business environment, where rapid information flow is extremely important, documents and messages are increasingly sent from one user, or system, to another over private or public *electronic-mail systems.*

Many proprietary electronic-mail systems are offered by the major computer vendors. For example, IBM has PROFS, Digital Equipment has All-In-One, and vendors such as Wang and Hewlett-Packard have their own. As with other proprietary technologies, these products are usually incompatible, and cannot communicate with each other without special interfaces being provided.

While the smaller manufacturers generally offer links up to the larger manufacturers' systems, the converse does not usually apply. For example, Wang supplies links to IBM and DEC electronic-mail systems, while neither IBM nor DEC supply links to Wang. Thus, the smaller the company providing the base system, the more links it must provide to other vendors' systems.

Since most large enterprises have a range of computer systems from different suppliers, a standard electronic messaging system which is available on all the machines is essential. Given this, many suppliers could supply products to meet those standards, on equal competitive terms.

The X.400 standard is the CCITT recommendation for an international electronic mail messaging standard. The ISO committee on message handling systems has ratified it and so it will become an ISO standard. Interim functional standards have been developed and are in use. As with other CCITT standards, there are four-yearly updates for X.400. ISO work on message-oriented text interchange systems (MOTIS) will build upon the X.400 CCITT/ISO standard.

X.400 is a 'store and forward' application, in which messages are sent to a mailbox for collection by a named user, who can access the mailbox from a range of terminals. The essential components of an X.400 system, shown in Figure 6.10, are a user agent (UA), and a message transfer agent (MTA). The UA provides services to the user, such as getting messages ready for sending in the correct format, retrieving and replying to messages, and the MTA carries messages from one UA to another.

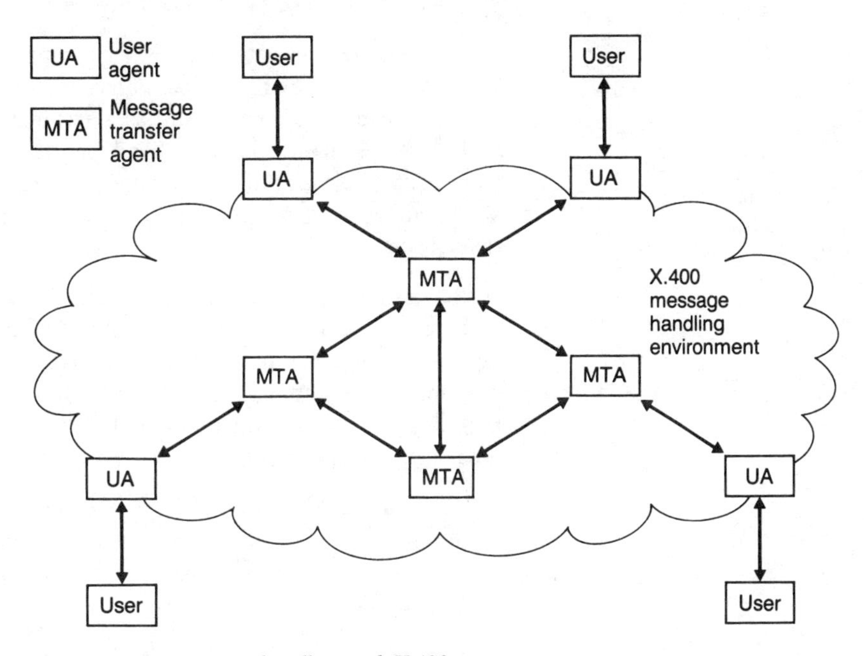

Figure 6.10 Message handling with X.400

The 1988 X.400 standards produced the message store (MS) definitions; the 1992 standards will include electronic data interchange (EDI). It is anticipated that there will eventually be one worldwide messaging system, able to handle electronic mail, images, telex, fax and digitized voice.

Directory management functions: X.500

The directory management functions on networks allow users of, or applications running on, OSI networks to obtain information relating to other users or applications on the same or other OSI networks anywhere in the world.

The base standards for the directory functions have been developed jointly by

ISO and CCITT. The CCITT recommendations, not yet ratified, are the X.500 standards. X.500 can be used within X.400 networks to provide a directory of the people who can be reached over the network.

Virtual terminal standards (VT)

With the move to open systems and subsequently, multi-vendor purchasing, a large number of terminals, each with characteristics that are proprietary to its own manufacturer, may be connected to a single system, which itself may be connected to a network. Since the variation in terminals may adversely affect the performance of the system, some systems have *terminal controllers* to take the load off the central processor. Software running on the network system must be able to recognize and deal with this collection of different terminals and terminal controllers.

In the past, differences in terminals from different manufacturers were not serious problems, because both the terminals and the controlling software were usually supplied by the manufacturer of the central system.

On a multi-vendor network, systems connected to the network may be accessed by any one of a number of terminals connected to the network, or by another computer system acting as a terminal. To run an application on a system, users input commands and data through the keyboard or mouse attached to their terminal or system and see the results on their screen. It should not be necessary for them to know where the application is running on the network.

Virtual terminal (VT) is a service which allows host systems to communicate with terminals and terminal controllers in a standard way over an OSI network. It assumes that the terminal has a certain amount of intelligence available in the terminal controller and that suitable software is installed on both the terminal controller and the host computer system. Standards for VT are in preliminary stages.

Work is proceeding to extend the VT protocols so that they will also be usable for communications between application programs. When this is achieved, it will be a major step towards the implementation of interoperability.

Job transfer and manipulation (JTM)

In complex networks, jobs often need to be submitted at one system on the network but run on another. Job transfer and manipulation (JTM) are the ISO standards emerging for this.

In the current definition of JTM, the user must specify which system is to run the job and must know the characteristics of that system. In the longer term, this should not be a requirement; scheduling should be done automatically by the system, and be invisible to the user. Eventually, it is hoped that JTM standards could lead to automatic sharing of tasks between systems on the network, a requirement for interoperability.

Functional OSI standards

Because the OSI scheme includes many possible choices designed to meet different user needs, it is left to groups of users to fill in the fine detail of the model, specifying precise packages of standards for their own industry application requirements. These *functional standards*, also known as *profiles* or stacks, can be specified to suppliers, who may then build products to support them. Users may be reasonably confident that equipment which meets these functional standards will be able to communicate easily.

Two large users, the US and UK governments, specified two of the early functional standards, referred to as government OSI profiles (GOSIP). Work is currently underway to *harmonize* (i.e. ensure the compatibility of) these with European Normes (ENs), functional standards specified by the European standards bodies, CEN/CENELEC.

The UKGOSIP is illustrated in Figure 6.11. This is the collection of OSI standards defined by the CCTA—the agency which guides UK government IT procurements—as suitable for UK government purchasing requirements. Other national governments have similar specifications.

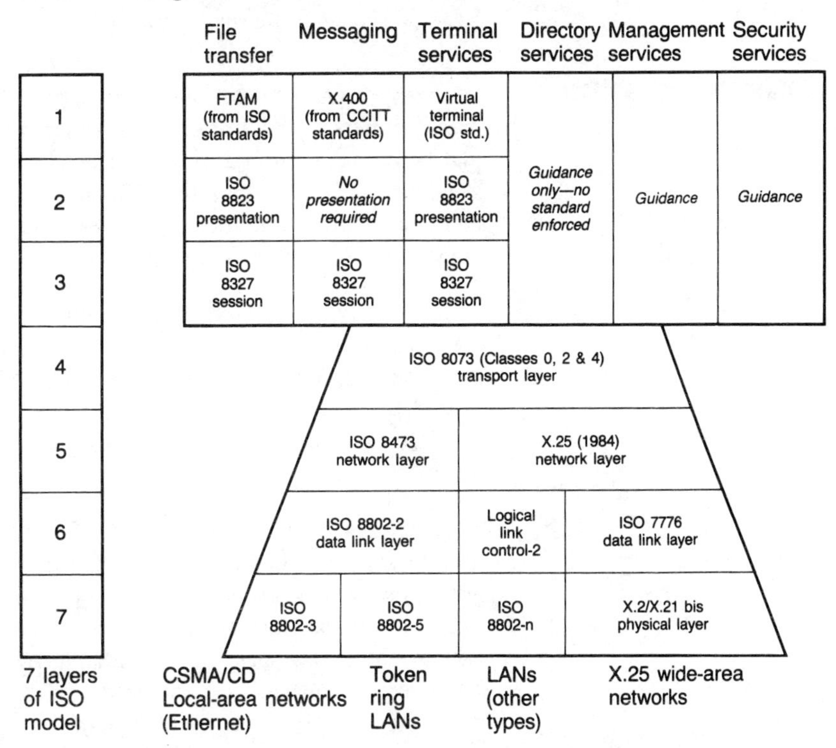

Figure 6.11 Standards for the seven-layer model—the UKGOSIP

The first important functional standard for commercial applications was the Manufacturers' Automation Protocol (MAP), specified by General Motors for its factory automation programme. Boeing subsequently added the closely related Technical Office Protocol (TOP).

Functional standards need to be harmonized as they are developed. ISO is developing a framework which will allow profiles to be compared, with the aim of preventing duplication and conflict in the specifications.

As well as creating new standards and harmonizing functional standards, ISO, in cooperation with many other standards organizations, has been carrying out work specifically aimed at making OSI standards more easily usable. Some of this is outlined below.

Network management

Standard communications for multi-vendor, un-alike systems solves some problems, but also creates new ones. One of the biggest of these is the management of a network of un-alike computer systems. To deal with this, information on the current activities of each layer of each system on the network will be needed at a central location.

ISO has agreed a structure in which information about the activity of all entities on the network is kept in an information database. The protocols by which this information is transmitted have also been agreed. The standard is sufficiently general to be adopted by the IEEE's POSIX committee as the basis of a standard for overall system administration.

Conformance testing

For many users, it is not only necessary to have standards defined, and for manufacturers to build products that meet them it is also necessary to have that conformance tested by an independent third party. As a consequence, standards activity within the OSI area has moved into the problems of defining tests for conformance. This has in turn led to the existence of test houses, which carry out suites of tests on OSI (and POSIX) implementations.

Unfortunately, such conformance testing only confirms that a product meets the paper specification of the standard. It is still possible that the product might not communicate with another implementation, even one which has passed the same test. This is partly due to variation allowed within the standard, and partly because the tests are still relatively new and contain inconsistencies.

Development work is going on to define the test procedures which can be used to guarantee that two OSI implementations will work together. In OSI terms, this is the meaning of interoperability. Because the ability for two systems to work fully together will be a function of the operating system on each machine as well as of the communications technology, the standards and standards-testing work on portability (POSIX) and that on interconnection (OSI) will need to converge at this point.

Additional standards for networked applications

Slowly but surely the world is showing a need for standards in very many applications areas, many of which are generated by the need for transmission of information. With the steady expansion of inter-company trading and the opening up of international trading, this is going to increase continually. As many of these requirements are closely tied to developments in OSI, we overview some of them here.

Office Document Architecture (ODA)

Increasingly, users are sending documents electronically across networks, from a machine using one office automation or business management package to another, which may be using different software. In order to avoid customized translation and conversion of the documents, there is a move to define standards for document specification. This would allow users to exchange documents electronically, retaining all the structural and formatting information contained in the document in the process.

The Office Document Architecture (ODA) standard, produced jointly by the CCITT and ISO, defines the basic structure of text in a document which may be sent by OSI systems. It allows a mixture of text and graphics to be contained in the document. Although the rules have been developed for, and are generally applied to, electronic documents, they are a formalization of the document structuring practices which have been used for conventional paper documents. Current work is focusing on the inclusion of colour. ODA formatted documents can be sent using X.400 or FTAM.

Electronic data interchange (EDI)

In the late 1980s, most companies began to realize the economic benefits of integrating systems both within their own organizations and across their suppliers' and customers' organizations in order to transfer information.

The information to be transferred can include for example, personnel data, mail messages, price lists, orders, quotations, invoices and inventory data. If this information is communicated electronically, companies can produce faster and better services for their customers. For example, if the product line in a factory is automatically coordinated with incoming orders, the company can carry less stock and produce to order.

Some needs are specific to particular industries: banks need to transfer funds securely, both within their own and between other banks; governments need to administer tax, social security, rates and other services through networks of distributed offices, using systems that communicate; manufacturing industries need to link machine tools on the factory floor with the raw materials inventory.

By the early 1990s, most businesses are expected to exchange information on orders, invoices and other documents, electronically. Manufacturers are already using such techniques to reduce their stockpiles of raw materials, moving to 'just-in-time' (JIT) manufacturing, where goods are ordered and supplied only as needed. Implementation of JIT requires well-designed, compatible systems at both manufacturer and supplier sites.

Many of the world's larger manufacturers of goods, such as automotive, aircraft and electronic companies, are already requiring compatible systems at their suppliers' and subcontractors' sites, in order that the necessary communications be as fast, straightforward, and error-free as possible.

Most of these applications of computer technology require international standards for electronic data interchange (EDI).

EDI standards concern themselves with definitions of common types of information and business documents, as used by agreed classes of organizations. EDI is generally considered as appropriate for the paperless exchange of information transferred between computer systems.

EDI generally moves into an organization in three stages. The first is simple: the replacement of paper by electronic storage. In certain organizations, this has become a massive problem. For example, the US Air Force takes an average of twenty months to review and distribute the two million pages of maintenance updates that contractors send out each year.

The second stage of the implementation of EDI within a company generally comes from recognition of the operational improvements created by reducing stock levels, achieving greater control over delivery timings, implementing just-in-time manufacturing and so on.

In general, it is estimated that more than 50 per cent of all computer output becomes input. For example, in many companies, data keyed in against price book information is processed and printed as an order which is posted to another company. There it is keyed in again as part of the order backlog. When the goods are shipped, an invoice is printed which goes to the customer. The customer receives goods and invoice and types the information into the computer again, to update the inventory. When the invoice is paid, the information is keyed into the suppliers system, and so on.

The high level of human input required to transfer information along the chain means that errors can be introduced and propagated at any stage. It is estimated that the use of EDI in such applications can cut costs by at least 10 per cent.

The third stage of the implementation of EDI is the most challenging, for it occurs when organizations start to use their corporate data and information to gain advantage over their competitors. This might be by gaining better control over the business overall, or to analyse the performance of supplier companies, or to better understand and supply to customers needs. This is the real motivation behind the current high interest in executive information systems (EIS) where the issues of EDI become managerial, rather than technical.

Use of EDI raises many related issues, such as the legality of transactions when performed electronically. For example, monetary transactions usually require signatures of authorized people in order to be legally binding. When is an electronically transferred signature legal? We are all familiar with automatic teller machines, where the signature as such does not exist. But what of document based transactions such as contracts, titles and deeds? The efficiencies to be had by quick, accurate and ratifiable transfer of documents are obvious.

Standards for EDI

Since the kind of information required to be sent varies from industry to industry, the standards work on EDI is necessarily widespread. Both the United Nations (UN), and the European Community (EC), have been instrumental in establishing EDI standards for specific industry sectors. This is a measure of their worldwide economic and political importance.

ISO EDI standards so far include definition of a dictionary of data elements for use in EDI communications, and descriptions of the systems in which they can be used. The EDI service should eventually run directly over X.400 and FTAM.

An EDI example: Computer-aided Acquisitions and Logistics Support (CALS)

The US Navy has 300 million microfilm cards of engineering drawings. Maintaining orderly records has become a serious problem, raising questions in the Department of Defense on whether military hardware is properly operated and maintained. To counteract this, the US Defense Department has mounted a $1bn program called 'Computer-aided Acquisitions and Logistics Support' (CALS) to create a partially paperless Pentagon.

From 1990, the CALS program requires all new weapons proposals, including engineering drawings, to be submitted on disks and tapes that Pentagon computers can read. Eventually, data will be sent electronically directly from the contractor's computers to those of the Services, eliminating disks and tapes. This program is expected to cut 20 per cent from the $5 bn the Pentagon currently spends annually on handling technical data for weapons.

The CALS standards cover virtually every aspects of documentation: text, illustrations, images and the way in which documents are transmitted. Graphics, such as digitized photographs, that can be stored by a series of lines of dots (rasters) are represented in CALS by the CCITT Group 4 Standard. This is also one of the standard formats for fax machines worldwide, Group III being the most common today.

CALS is a two-phased initiative. Phase 1 specifies the standards which will be used to interchange documents between heterogeneous computer systems. The standard specifies text, graphics and communications standards, all of which are based on international computer industry standards. By 1990, aerospace and defence contractors will begin submitting weapon system documentation to the DoD in electronic form. This will obviously invoke requirements for highly secure systems.

To implement the CALS program, standards were first set for computerized communications between contractors and subcontractors. Ultimately, in the 1990s, the Pentagon expects to be able to access its contractors' computers, to view designs in progress and look at technical information. If a supplier runs into trouble, the information will then be immediately made available to other organizations.

Although government organizations are generally the most dramatic examples one can find of huge users of paper, some private industries are not far behind. Realizing 'paperless' databanks of shared information will not be achieved easily without the use of far-reaching and generalized industry standards designed to take some of the variability out of the basic system components.

Because of the widespread and universal nature of the CALS standards, they are expected to be taken up by many other organizations around the world.

Integrated Services Digital Network Services (ISDN)

Complementary to the OSI standards are those for the Integrated Services Digital Network (ISDN). This network is designed to enable the transfer of very large amounts of data quickly to anywhere on a worldwide network. ISDN accommodates a wide range of services over the same line, including voice, data, videotext, fax and video telephony.

In 1989 Europe's PTTs agreed the standards for a pan-European ISDN, the specification of which will allow users to invest in equipment with an assurance that it will not become redundant in the future. There are three requirements in the agreement. First, the PTTs will provide a common range of services, to a common standard. Second, each will support common standards for customer equipment, so that these will work in any country that is party to the agreement. Third, the national services will be interconnected to produce an international set of services.

The agreement has been signed by 18 countries, representing over 350 million people.

The UK, France and Germany all provide ISDN to major organizations at present, but not to a common standard. Standards for customer equipment will be handled by the European Telecommunications Standards Institute, ETSI, set up in 1988. The Institute includes national telecommunications administration operators, industrial representatives, users and research bodies.

Asian interest in ISDN is increasing steadily, with the Asian ISDN Council (AIC) formed in 1988. The Asia-Pacific Telecommunity (APT) is working to raise the level of Asian telecommunications through education and other information-sharing activities. ISDN is seen by many participants to be key in achieving and retaining Far Eastern competitive advantage in many industries.

OSI and ISDN

In June 1989, the US Air Force conducted what is believed to be the first network trial combining OSI and ISDN protocols. The trial was initiated by US NIST and features OSI messaging and file transfer applications running over the US Air Force's existing network, based on Integrated Service Digital Network (ISDN) technology. AT&T managed the project.

The trial demonstrated ISDN as both an inter-networking vehicle, linking both local and wide area networks, and as the underlying network technology for OSI applications. Although the trial was only for data communications, an integrated OSI-ISDN environment has the flexibility also to handle voice and video. This will be very important in the future.

The integration of OSI and ISDN technologies is important to the US government because ISDN is expected to be included in upcoming versions of the US GOSIP specification. Federal regulation mandates that US government then use the specifications in their telecommunications procurements. The first version of US GOSIP, released in 1988, will take effect in 1990 and the second version, expected to include ISDN specifications, will take effect in 1991.

Interoperability

Interoperability is the ability to have computers from different vendors work together in a truly cooperative way over a network. This means that the machines must interconnect intelligently and that applications running on different machines work closely together, sharing tasks as appropriate.

To achieve interoperability, standardization work on the operating system interface and application environment on a single machine (POSIX) must be merged with standards work on interconnection (OSI). Then networks can be used not only for sending information and commands but for sharing processing tasks.

Much of the work going on in this area centres on defining remote procedure calls (RPC) which request processes on remote machines as if they were local.

The emerging *de facto* standard for RPC is Network Computing Architecture, (NCA), from Apollo, now part of Hewlett-Packard. NCA has been licensed to a number of other manufacturers, many of whom have committed resources to the Network Computing Forum (NCF) to cooperate on further development of NCA.

In the UK, the ANSA (Advanced Network Systems Architecture) project, sponsored by a government research initiative known as Alvey, has been investigating ways to produce generalized distributed systems. To date this is a research project only.

ISO is defining a reference model for open distributed processing (ODP), to be a framework of the same nature as the OSI Seven-Layer Model. This model is expected to draw on the research work of ANSA and of the co-operative European IT research program known as ESPRIT. The specific parts of the ESPRIT program are the Computer Networks for Manufacturing Automation (CNMA) project and the Computers for Integrated Manufacturing—Open Systems Architecture (CIM—OSA) projects.

Summary: standards for interconnection

In this chapter, the international OSI standards for interconnection have been reviewed. These are supported in principle by the entire computer industry, but there are few products yet that implement them fully.

For practical use of the OSI standards, there are many related standards that need to be specified. Some of these are applicable to all industries, while others are specific to certain markets. Work is proceeding to define many of these.

The driving force behind much of the standardization work in communications is the increasing need for electronic trading between departments, companies and countries. The global market demands new, high-speed, international networked systems offering accurate and fast communications. These must be designed to support the multi-media systems of the future, an the variety of hardware in common use today.

If we are to deal with the applications problems of the r 1990s, we must 'make IT easier' by removing some of the exists today at the system level—both hardware and softwa

this by using more products that conform to internationally approved, widely supplied and supported standards.

Without that, valuable resources of both time and creativity will be continually squandered on making incompatible sytems work in harmony, rather than on solving the applications problems of the 1990s.

CHAPTER 7 The role of X/Open

In previous chapters, we have discussed the general need for standards in the IT industry. We identified the standards that are required today, and the general means by which standards are set. In particular, we surveyed the current and emerging standards for portability, scalability and interoperability, together with some extensions that are particular to certain applications. As the picture has unfolded, it should be clear that the area of standardization in the computer industry is vast, difficult, vitally important, and potentially confused and confusing.

In this chapter we look at X/OPEN—an organization that is attempting, in collaboration with many others, to bring structure and discipline to the overall area of open-systems standards. X/Open is beginning to succeed in its principal objective—that of bringing open-system standards under one comprehensive umbrella—and the standards which it is adopting and promoting are slowly but surely gaining worldwide, and widespread, endorsement.

The history of X/Open

X/Open was founded in 1984. It was set up as an international non-profit organization by a group of major computer manufacturers 'to facilitate, guide and manage the process that will allow commercial and government users, software vendors, standards organizations and systems makers to solve the information technology dilemmas caused by incompatibilities in their systems and software components'.

Today (1990), most X/Open shareholders are major computer manufacturers. Among these are: AT&T, Bull, Digital Equipment, Fujitsu, Hewlett-Packard, Hitachi, IBM, ICL, NCR, NEC, Nixdorf, Nokia Data, Olivetti, Philips, Prime, Siemens, Sun Microsystems and Unisys. There are two additional members from the industry associations, both of whom joined in 1989. These are: the Open Software Foundation (OSF) and Unix International (UI). Other contributors to X/OPEN activities represent users, independent software vendors, system integrators and government agencies, so balancing the interests of the other constituents of the computer industry.

A requirement of membership of X/Open is a serious commitment to ship products that support, or conform to, the IT standards which X/Open defines.

X/Open, which was founded by the major, indigenous European manufacturers, has been an international organization from the start. As its membership expanded to include major US and Japanese manufacturers, so its interests and mission have become truly worldwide.

The X/Open mission

'... to bring greater value to users from computing through the practical implementation of open systems'.

To achieve their mission, X/Open is coordinating the development of a Common Applications Environment (CAE), which is being supplied on computer systems by both its own members and by other manufacturers. The CAE comprises all the components of a computer that need to be specified and standardized if the characteristics of portability, scalability and interoperability are to be achieved.

In order to define the CAE, individual standards for the various components are adopted and adapted by X/Open from existing public or *de facto* standards, as appropriate. These are then moulded into one cohesive and comprehensive superstandard (see Figure 7.1).

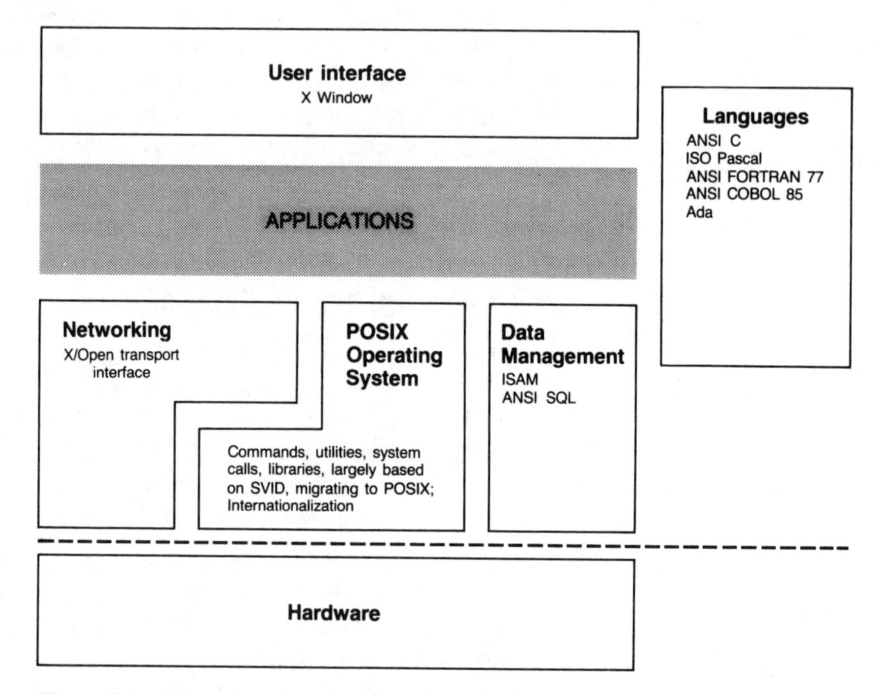

Figure 7.1 X/Open's common applications environment

The goal of the CAE is to provide application software portability by defining the following three principal requirements:

- a common user interface;
- a common programmer interface;
- a common connectivity model.

All three requirements are supported by hardware systems to be supplied from a wide variety of manufacturers.

Volume shipments of computers that support the CAE would address many of the issues raised in previous chapters, for there would then be:

- greater choice in system vendor selection;
- improved, and simplified multi-vendor system integration;
- lower training costs for personnel as the usage of the CAE becomes widespread;
- longer product life cycles as investments in software and know-how become, at least partially, protected over time;
- lower technical support costs for all links in the software and hardware supply chain, due to the reduced variability.

In other words, X/Open, with its CAE, is aiming to address all the issues that are the subject of this book.

If the CAE were truly to incorporate the best of all available work in the standards area, and if it were promoted sufficiently so as to reach its potentially high levels of uptake, the return on information technology investments could at last improve. Organizations' resources could then be spent on positive, growth-oriented, practical solutions to the real problems of their own businesses.

As discussed in Chapter 2, software developers specifically could benefit from the compatible and growing market that widespread use of the CAE would produce. They would have a large and compatible user base to target and could therefore spread their software development investments over a large base of machines, without a corresponding growth in the support and development costs.

Increasing the number and variety of applications available for the open-systems environment should spur computer market growth overall. With fragmentation reduced by use of common standards, a large pool of applications solutions could exist, and users could choose appropriately. The result could be a growing market for CAE-compliant hardware systems and heightened opportunities for vendors of both hardware and software.

The Common Applications Environment

The X/Open common applications environment (CAE) is designed to provide a complete support system for the development and running of application software. It is designed to enable high levels of portability, interconnection and interoperability across machines.

To satisfy its objectives, the CAE defines common applications programming interfaces to the many services required by a sophisticated application. These basically constitute the following:

- a common syntax for the major programming languages;
- a common approach to interfacing with the user;
- a sophisticated approach to multi-vendor networking.

Because definition of the complete CAE is a huge and constantly evolving task, it is under development in stages. Each stage is designed to enhance the portability of applications significantly. The first two steps provided definitions of:

- the interface to the operating system;
- the data management services;
- the C, COBOL, FORTRAN, Pascal, and Ada software programming languages;
- common media standards to facilitate the transfer of source code between systems.

The third step, published in 1988, added common networking and window management interfaces, as well as additional language functionality.

The definitions comprising the CAE are contained in a set of books describing each standard and available from book suppliers worldwide. These books are collectively known as the X/Open Portability Guide (XPG).

If application developers utilize only those interfaces which are defined in the XPG, they can ensure that their applications are totally portable at the source code level. Specifications currently defined in the third edition of the Portability Guide, known as XPG3, are described below.

Operating system services

The basis of the CAE is a common set of interfaces to the operating system services. These are based on the UNIX operating system interfaces, but have been modified and extensively annotated. There are no dependencies on the features of any proprietary implementation of UNIX, nor on any specific hardware architecture.

Since the policy of X/Open is that it will always adopt public standards from official standards bodies if available, it fully supports the IEEE POSIX standards, and will continue to do so as these evolve.

It is vital to X/Open members that the systems which they supply have the ability to cope with a variety of natural languages and cultural conventions. XPG3 therefore defines a comprehensive set of internationalization standards specifications. In this area, it is often working ahead of the public standards bodies.

Programming languages

As we saw in Chapter 5, the C language is fundamental to standards work on the operating system interface. It is, therefore, also fundamental to the definition of the X/Open environment and to the definition of the CAE operating system interface.

The X/Open definition of C is based on the ANSI 1989 standard. Over time, the X/Open C language definitions will be modified to align fully with the emerging ISO standard.

The XPG defines four major high-level languages commonly used in both open and proprietary system environments. Definitions for FORTRAN and Pascal are the widely adopted ANSI FORTRAN-77 and ISO Pascal, standards respectively. The definition for COBOL is the ANSI COBOL 85 standard, extended to support interactive operation. XPG3 has also been supplemented to include the Ada language, as defined by the US Department of Defense.

Data management

X/Open has adopted common interfaces for the creation and management of files and for access to relational database systems. For access to relational databases, an extended subset of the standard ANSI database language SQL is defined in the CAE. This supports embedded SQL calls within C or COBOL source programs.

Networking

X/Open initially concentrated its efforts and resources on the issues relevant to portability and scalability. More recently, responding to the growing importance of distributed applications, X/Open is beginning to address networking and interworking more aggressively, in collaboration with the networking standards groups.

Here, it will clearly need to merge its work with that of ISO and related organizations, which have been working for many years on the definitions of the OSI standards.

As was mentioned in Chapter 6, the X/Open transport interface (XTI) will define a programming interface to the OSI transport (Level 4) services. This interface is totally independent of underlying protocols and can therefore be used with ISO protocols or with other widely used protocols, such as TCP/IP.

Window management

A quality user interface common to many different applications, is a high-priority requirement for many users, who see it as an important ingredient of 'people portability'.

X/Open is defining a complete user interface to the *de facto* standard, the X Windows networked window management system. The interface is being jointly developed by MIT and a consortium of systems vendors.

With the possibility that more than one *de facto* standard for user interfaces may emerge in the marketplace, specifically Open Look from AT&T and Motif from OSF, there may well be a requirement on X/Open to define its CAE in such a way as to allow the option of using either of these, or indeed, any other that meets the specification.

Security

As we have seen, the security of operating systems is an important element in many commercial data processing environments. X/Open's CAE offers security features that are consistent with emerging standards. It has developed a handbook defining these security features, including explanations of how to utilize them fully.

Support for X/Open and the CAE

The standards that X/Open is using in its CAE are all taken from existing, widely supported, public and *de facto* standards. The job of the technical team within X/Open is to adopt and adapt those standards, merging them into the CAE, while that of the members is to design and market machines to support it.

But the work going on inside X/Open represents only one part of the work required for progressing open-systems specification. Users need to take an active part up-front by following the issues and providing input to the requirements process. They then need to specify systems that comply with X/Open's CAE in their procurement guidelines and to develop and use applications that meet those specifications.

X/Open's efforts to establish an open CAE currently have the publicly stated support of NIST in the US, together with that of the European Community, EC, and of many other government organizations worldwide. X/Open's current CAE guidelines have been adopted within the procurement policies of key European and other national governments, as well as by several United States multinational corporations and government agencies.

Hardware and software vendors need to use the X/Open Portability Guide as a reference for application and system development and to ship, in volume, products that meet the X/Open specifications. Today, many hardware manufacturers are shipping hardware that supports the CAE, albeit in low volumes as yet, and often to a minimal subset of features. Similarly, several hundred software developers have declared their intention to build and supply CAE-conformant products.

X/Open, in conjunction with the international users association Uniforum, publishes a catalogue of applications software products that conform to its specifications. It also publishes a bi-monthly newsletter designed to keep the industry informed of developments related to the CAE.

X/Open's sponsors committed themselves to make open systems a practical reality when they joined the organization. However, they will only be able to achieve this in practice if they have the full support and constructive help of all other elements of the computer market.

The X/Open technical process

The technical achievements of X/Open in its relatively few years of existence are the results of a complex process that employs the technical expertise and commitment of the organization's member companies, together with that of hundreds of individuals worldwide.

Each year, X/Open conducts approximately 80 meetings with technical experts from its member companies. These include meetings to establish and manage the organization's overall technical process, and meetings to work through the details of future X/Open portability definitions. These meetings are supported by an extensive work program that is carried out both by the skilled staffs of member companies and by expert subcontractors. In total, some 200 men and women contribute regularly to the X/Open technical program.

The technical meetings are held around the world, in different locations in rotation. Although this imposes a heavy travel burden on participants, it ensures that the load is shared by all participants. It also brings X/Open into contact on a regular basis with various parts of participating companies in different areas of the world, thus contributing to the international flavour of the work.

Defining the process

The technical process within X/Open starts with the X/Open board of directors. They have the responsibility of establishing the overall strategy for the activities of X/Open.

Each member company holds one seat on the board of directors. In addition, each member company provides a technical manager to the organization. Since these people are responsible for representing their companies' technical viewpoint within X/Open, and for conveying the reasons for the decisions taken back to their own companies, they must be senior, highly skilled technical people. They also carry a certain amount of responsibility for promotion of X/Open's aims and objectives within their own companies.

It is the technical managers group that must turn the board's strategic decisions into an executable technical work program.

The technical managers usually convene at least eight times each year. During these meetings, they review the implications of the ever-changing market in which X/Open operates, and they aim to resolve the inevitable differences of opinion that arise among representatives of so diverse, and competitive, a group of companies. Once each year, the technical managers group assembles for a full week to prepare recommendations for components of the forthcoming year's technical program.

The bulk of the technical work is carried out by the organization's technical working groups. These committees meet at regular intervals to review progress, resolve differences and recommend courses of action for the forthcoming period.

Supervision of each of the technical working groups is assigned to one of the members of the technical managers group. Some technical managers personally manage the day-to-day activities of their working groups; others delegate the task to one of their own staff members.

About the working groups

The Kernel Group is one of the most active of the X/Open working groups, meeting about twelve times annually. Its main job is to ensure that the work of the IEEE POSIX working groups and that of X/Open are converging.

The Kernel Group is responsible for making any changes to the CAE that prove necessary to ensure that the X/Open operating system interfaces are fully compatible with the emerging POSIX standards.

Working alongside the Kernel Group is the Command Group. This is working closely with the IEEE on an improved definition of operating system commands and interfaces.

The Distributed Processing Group and the PC Interworking Group focus on the interworking between different, not necessarily open, systems. This is a natural and essential extension of the activities of X/Open—to help users incorporate open systems into their networks without sacrificing the value they currently realize from existing, proprietary systems.

The Security Working Group is addressing the growing demands, both from government and commercial users, for increased protection of data and systems from unauthorized access. The first results of this work were the production of a comprehensive guide to the security features currently available in X/Open systems,

The User Interface Working Group is responsible for X/Open activities in the area of user interaction. The emphasis is on the use of high-quality graphical user interface techniques, which are available on modern bit-mapped screens. The first phase of this work was a definition of application interfaces to a networked window management system, X-Windows.

In the longer term, this group expects to define a policy for the CAE in the controversial area of the 'look and feel' part of the user interface. This is a very difficult area, both politically and technically, for the users' demands for interface consistency between systems is in conflict with the desires of those vendors wishing to make this an area of proprietary added-value.

X/Open has always recognized that it is not only important to build products that meet defined standards, but it is also vital to be able to demonstrate such conformance. A very important aspect of X/Open's program is, therefore, the development of a test suite that will be used to verify compliance to the CAE. The Verification Working Group has been responsible for development of an extensive test suite that is available to members and non-members alike as the basis for the branding of compliant systems with the X/Open trademark.

The X/Open verification and branding program

X/Open has defined three levels of CAE compliance for computer systems, and one for application software. The first, X/Open Base, is the minimum system brand. It covers hardware and the base-level components of the CAE required to ensure that the system supports a comprehensive operating environment. The second, X/Open Plus, identifies a computer system that supports the complete X/Open CAE, with the possible exception of one or two components which have been designated as optional. The third brand, X/Open Component, may be applied to individual conforming software components of the CAE. The fourth brand is X/Open Application, designed for software applications that meet the CAE documentation requirements.

To receive the X/Open brand, computer hardware vendors must enter into a trademark licence agreement with X/Open. Their computer systems must successfully complete the X/Open verification suite in a single run, with no modifications and no errors. The systems must also satisfy referenced tests from US government agencies such as the National Institute of Standards and Technology (NIST) and the General Services Administration (GSA).

Software applications must meet a stringent checklist of X/Open specifications in order to be declared compliant. The XPG3 version of the verification suite, for example, consists of approximately 4200 tests for the C language, operating system, internationalization and data management implementations.

Branded products are identified by X/Open logos. Each logo is accompanied by a box indicating the nature of the compliance and the version of the Portability Guide against which compliance is claimed.

Tests for hardware compliance can be conducted at various approved locations in the US, Europe and Japan, with additional sites planned. If they wish, X/Open member companies may conduct the verification suite process for themselves. All companies—X/Open members and non-members—will eventually be able to self-test, but only after X/Open issues an updated version of its test suite.

Under the branding program, computer hardware vendors pay royalty fees to X/Open against shipments of branded products. In addition, product and licensee registration fees are applied; these vary from country to country according to national trademark laws. There are no royalties payable on shipments of software applications which carry the X/Open brand.

Completing the cycle

Completion of a particular phase of the CAE and acceptance of its recommendations by X/Open is a complex process.

First, it is the role of the working group to produce a detailed definition of a component in its area that has the broad support of the participants. During this phase, any major areas of disagreement should be resolved at the technical manager's level. The board of directors will intercede during this phase, if necessary.

As the various definitions are agreed upon, the X/Open professional staff takes responsibility for the editorial integration of the various components into a coherent, consistently presented document.

When complete, the new version of the Portability Guide is submitted to the board of directors for approval. During the development, great care is taken to ensure that this final approval is merely a formality—that all issues, no matter how large or small, have been resolved already.

X/Open is structured so that no single company or minority group should be able to exert undue influence or veto publication, however, publication requires broad consensus. Once the Portability Guide has been published, all member companies agree to support it with compliant products.

X/Open then distributes and promotes the Portability Guide worldwide, to ensure the broadest possible exposure. Its objective is to encourage the development of products that conform to the latest CAE specification and to persuade purchasers of products to specify X/Open conformance in their procurement policies.

This can be a long and difficult process in its own right, not least because many organizations already have rules and regulations governing procurement. For example, the European Commission cannot use in its procurements, standards that have not yet been adopted by the European standards bodies, CEN/CENELEC. Therefore, it can have problems with those parts of the CAE that reference standards which to date only have US ANSI approval. Thus the entire process tends to be an iterative one, with each stage bringing all parties closer to universally approved standards.

User and ISV advisory councils

By defining the CAE, X/Open can clearly set the stage for practical implementation of open systems. Its members can carry the process one stage further by shipping the systems that support it into the marketplace. But it is the users and independent software vendors who will build the conformant applications software and buy the products, and it is towards these groups that X/Open members are directing their open-systems solutions.

To help set its future technical and marketing priorities, X/Open formed permanent user and independent software vendor (ISV) advisory councils in 1988. These Councils, which meet on a regular basis, are composed of senior executives from large user organizations and software firms in both the US and Europe.

Corporate membership of the X/Open ISV and user advisory councils, as of February 1989, are shown in Figure 7.2.

To emphasize the importance of the role of these councils, the board of directors of X/Open now includes representatives from both.

The X/Open Software Partners Program

In 1988 the Software Partners Program was introduced. This now has more than 200 participants worldwide.

The program supports independent software vendors in two key ways. It provides information to help them build products that are conformant to X/Open standards, and it helps them to reach the open-systems market.

Program services include:

- the verification and branding program;
- direct access to work items developed by the various X/Open technical committees;
- the X/Open Software Directory, which profiles and promotes the products of Software Partners Program members;
- copies of the XPG;
- a subscription to the X/Open newsletter, *Open Comments*.

User Council	ISV Council	System Vendor Council
Amoco Corp. (US)	ASCII Corp. (Japan)	ARIX Corp. (US)
Arco Oil & Gas (US)	Informix Software (US)	Omron Corp. (Japan)
Automobile Association (UK)	Ingres Corp. (US)	Pyramid Technology Corp. (US)
Bellcore (US)	Interactive Systems Corp. (US)	Sequent Computer Systems Inc. (US)
British Telecom (UK)	Liant Software Corp. (US)	Sony Corp. (Japan)
Central Computer & Telecommunications Agency (UK)	Micro Focus (US)	
Commission of European Communities (Belgium)	Microsoft Corp. (US)	
Daimler-Benz AG (West Germany)	Novell, Inc. (US)	
DHL International (US)	Oracle Corp. (US)	
DuPont (US)	Progress Software (US)	
Eastman-Kodak Co. (US)	Quadration (US)	
Elf Aquitane (France)	The Santa Cruz Operation (US)	
Exxon Production Research (US)	Softlab GmbH (West Germany)	
Ford Motor Company (US)	Sybase (US)	
Gerling Konzdrn (West Germany)	Tecsiel-IRI Finsiel (Italty)	
Harris Corp. (US)	Tietotehdas Group (Finland)	
Hoescht (West Germany)	Unify Corp. (US)	
Landesamt fur Datenverabeitung und Statistik (West Germany)	Uniplex, Inc. (UK)	
McDonnell-Douglas (US)		
National Institute of Standards & Technology (US)		
Shell International Petroleum (The Netherlands)		
Swedish Agency for Administrative Development (Sweden)		
Union Bank of Switzerland (Switzerland)		

Figure 7.2 X/Open advisory council

The X/Open Requirements Conference (XTRA)

The success of any standards related group can be measured not only by the influence it has in the marketplace, but also by the enthusiasm and breadth of its support.

The pre-eminent position of X/Open in providing a truly representative role was clearly demonstrated when in 1989, X/Open held the first of its planned Requirements conferences (XTRA) in Montreal, Canada. Entitled 'Shaping the Future of Open Systems', its objective was to bring together interested parties to discuss and prioritize the work required to bring the potential for open system to a practical reality.

The conference, which was attended by a large number of users, ISVs, system integrators, government agency representatives and hardware vendors, was organized into working groups, each of which addressed a specific functional area (see Figure 7.3).

- Open systems' directions and migration strategy
- Operating system environment
- Data management
- Application development tools and languages
- Internationalization
- Networking and communications
- Security
- Systems administration and management
- Human computer interface

Figure 7.3 Requirements Conference (1989)—working groups

At the end of the conference, the results of the working groups were presented to the complete group, and the whole group asked to express their view on the relative priority of the requirements identified. An analysis of the ratings, which will be published in detail by X/Open, provides valuable insight into the relative priority of each requirement, and to the different priorities of the various groups represented.

This conference, which will be repeated regularly at various locations, will serve to guide the research and development priorities within X/Open, and gives a powerful way for representatives of the open systems purchasing community to communicate their needs directly to the organization. Full results of the May 1990 XTRA will be published at the end of 1990.

In the next chapter we will report on the results of this first requirements conference.

Visions of the open decade

We finish this chapter with a statement from the president and chief executive of X/Open, Geoffrey Morris, as given in the keynote presentation at the 1990 Uniforum Conference, held in Washington DC.

The presentation details the X/Open perspective of the future of the open systems marketplace and shows their commitment to a united and standards driven industry.

The UNIX community, which gave life to the Open Systems community, is shaping a future information-based economy. Now, without partnership, we will not be able to meet the mighty challenges that face us over the next decade—the 1990s—what we at X/Open call *The Open Decade*.

It is that vision—our Vision of the Open Decade of the 1990s that I want to tell you about today—I want to tell you about what we believe we can achieve through active partnership by the year 2000, I want to tell you about what we need to do to create the open platforms that will get us there, and I want to tell you about a process we call *XTRA* that will make sure those platforms evolve fast and effectively.

Now, I am going to talk about the future because that's where I expect to spend the rest of my life.

Legacy of the past

But first let me describe where we are now. The eighties was a period where I thought that our industry had lost some of its interest in the future—it did not seem very adept at looking more than a couple of years ahead of itself—and sometimes no further than the next quarter's financial results on Wall Street.

That went for many customers as well. They wrestled—they wrestled with all kinds of alligators—they wrestled with personal computers—they wrestled with end-users' needs—they wrestled with incompatible systems—they wrestled with how to squeeze more out of their budgets—they wrestled with how to reposition their MIS departments in the corporate structure. Some just sat back and let the suppliers determine the future of their corporate information systems.

Unfortunately the most tragic victim of this process was information—the very central commodity of what we've called the information economy that lies ahead of us in the next few years.

The information ghettos

In fact much of it never got to the stage of becoming information—it is still data—raw uncoordinated data—and lots of it—often under-used and trapped in some technological ghetto in desperate need of liberation—trapped in the closed and isolated information ghettos of government and corporate systems around the world.

This has got to be the worst legacy of forty years of the computer industry, and after an investment of over a trillion dollars.

We have *got* to get out of those information ghettos—and got to get rid of the constraints. Up to now people have been able to ask questions of information systems, but they seldom have been able to get the right answers—or even indeed an answer that's close to their question. Tomorrow has got to be the age of *the answer*.

We must bring the data out of those dark vaults of incompatibility, those crypts filled with data, that has accumulated as the residue of years of closed systems. We have got to provide platforms on which we can turn that data into information. And we must open the way for people to receive and use the *right* information through *any media* that suits them.

Those are the challenges for the 1990s. That's the challenge for the open systems movement. That's the X/Open vision for the year 2000.

The future lies with those who most successfully shake off the constraints of the past. In an important sense, the history of true information systems starts now.

Outlook 2000

Let me try and put some context to the end-state that we can expect by the year 2000 and to some of the reasons why.

Today in the computer industry we have a preoccupation with organizing information for users in government and business. Likewise the telecommunications industry has a preoccupation with how people communicate to each other. The broadcasting industry and the publishing industries have a preoccupation with the best ways of presenting organized information to people.

X/Open believes that the year 2000 will see the successful information providers being able to combine the organization, the communication and the presentation of information into one system which, of course, will be truly portable wherever you are—at home, or perhaps even in the transporter.

The combined demand and potential revenues for this information industry are impressive, and the business and economic arguments for an integrated information industry are clear. Today's growth in radio, telephone and cellular communication, television and computers is a strong message that there are billions of individuals who want answers—on the spot—and in a form that makes them meaningful.

So when we are sitting here on a January morning in the year 2000 we will not be talking about UNIX, we will not be talking about open operating systems, we will not be talking about how the information management systems of the year 2000 utilize the open technologies like broadcast, TV, film and digital publishing techniques.

The technologies that will support this future integration of information are already underway—cellular communications, ISDN, high definition television and VDI—particularly developments in television that by year 2000 will make today's intelligent workstations look like yesterday's dumb terminals.

These technologies can only be integrated if *interfaces* are developed to link them together. Developing those interfaces is essential to the future of the industry and the information economy of the year 2000.

Many corporations are already making the right kind of moves in the direction of producing a kind of multi-media cellular 'work-station'—a sort of futuristic voice, picture, text and music mixer. In their various forms these new work systems will support rapid and easy access to a vast range of valuable, multi-media, knowledge bases of information.

It is a future of personalized and interpersonal computing, of voice activated visual technologies and totally mobile personal systems—a global, mass market, a consumer information industry.

I think the single most important thing to understand about the prophecy is that in the information technology sectors, the requirements and the capabilities of products in the year 2000 will be driven by the needs and the purchasing power of the ordinary person—not just the computer scientist. The rising power of the customer is driving the industry towards this future.

So how is all this innovation occurring and why should I be talking in these terms at what is effectively a UNIX conference? It is because UNIX has accelerated the world towards an open future.

The open decade

At X/Open, we believe that the key to that vision of the future is the development of truly open systems. They provide a real opportunity for a new industry to develop and for the user community to influence and benefit from the progress.

There has got to be a new industry built on a solid platform of agreed industry standards that provide the bedrock of confidence, security and innovation for the future. The new industry will *have* to be based on standards to allow user organizations to plan ahead and create information networks that serve a variety of widely dispersed users.

If those standards platforms are not created, we will never integrate those key technologies, we will never be able to get the best out of those sophisticated knowledge bases, we will never be able to escape from the *information ghettos* and we will never realize the value of what we have at our fingertips.

For the 1990s to become the Open Decade there is a lot of work to be done. The way to achieving those open systems is a lot clearer right now than it was five years ago when X/Open began—and the key to that process is value—value for people who use technology. That is what X/Open is all about.

The X/Open mission

....Bringing value to users through the practical implementation of Open Systems.

We have created a structure for X/Open that can provide value across all sectors of the information technology industry. X/Open is at the heart of the practical Open Systems process. The industry at large, system vendors, software vendors—and most importantly users—can work together with our members through our different councils to provide X/Open with a broad cross-section of views, skills and needs for Open Systems.

We are pulling all these sectors together in a partnership that creates the vendor independent—and product independent—standards specifications that are needed to make open systems a reality. These open systems specifications are documented in the XPG, a Portability Guide which together with conformance tests and branding tests, help users know that a product is XPG compliant.

X/Open members

The X/Open partnership has grown rapidly over the last five years, starting with hardware vendors, now including the 20 largest suppliers in the world, as well as software companies and users.

The participation of these users is crucial to our understanding of what is needed from open systems, to make them a practical and valuable reality.

And the collaboration is *working*. The evolving standards published in the X/Open XPG portability guide are becoming crucial planning tools for those organizations who want an open decade. Among some of the world's major user organizations that are specifying X/Open compliance are Harris Corporation, Bellcore, DHL and the US Treasury.

On Monday this week (January 22 1990) the West German Federal Government announced that all the 23 ministries—and the national railway system plus the Bundesbank—will base their purchases on X/Open compliant products.

And how do you identify the open systems products? Look for the X/Open brand is the answer.

The X/Open partnership has shown that it can work and that it can add real value to the industry. Certainly the industry has changed as a result. Our success in fostering and specifying an industry-wide platform, the XPG environment, has shown that we can converge standards on one platform.

But that's only the beginning. Too many people still think that open systems are the same thing as an open operating system . . . and they are not.

What we need next is an environment that can be used to build information systems of real value for business and government, to realize the full potential of the information application—which takes the full opportunity of innovative technology. That is the open systems vision. It is certainly the X/Open vision.

Today information technology is an integral part of our life and economy. Technology costs in industrialized nations represent at least 5 per cent of the value of the economic activities worldwide. Who forecast that at the beginning of the 1980s? Who even guessed the names of the companies that would make a difference in the information technology industry during the 1980s?

When we began the eighties the Personal Computer was just coming out of the garage, and using technology as a competitive weapon had more to do with Space Invaders than portable computers in the Board Room. Back then names like Compaq, Sun, Oracle and MIPS—and also X/Open—did not even exist. But what they all did was to realize the potential of standards as a way ahead.

The backdrop to this process is that the technological stampede in the industry has not been matched by an overall increase in purchases over the decade. In fact, for some years, vendor sales have barely kept pace with economic growth, and profit margins are in decline. Industry growth had simply slowed down—the percentage increase in US computer manufacturer sales between 1983 and 1988 is down to around 10 per cent from over 30 per cent at the beginning of the decade.

Market maturity—not yet

No corporation seems to be immune. It is not a company issue, it is an industry issue we are facing. Many analysts speculate that the industry issue is one of maturity—I cannot believe that. In my book market maturity means slow or no growth markets where products have penetrated to near saturation. How can we pin this label on an industry where only 20 per cent of the management professionals have access to desktop information systems? That indicates an 80 per cent opportunity.

So, we have to work out ways of overcoming the obstacles to growth. It has often been said that civilization advances by extending the number of important things that we can perform without thinking of them. Standards let you do just that because industry standardization results in a reduction in unnecessary differences and unnecessary differences that defeat compatibility are very costly *and* have no value.

Standards make markets

It is interesting how suppliers have put away their differences and how standardization has stimulated market growth. This is how it works. Figure 7.4 illustrates the shake out in vendor standards for a product as *de facto* industry standards emerge. The lower graph shows how the market starts to grow once the shake out has happened.

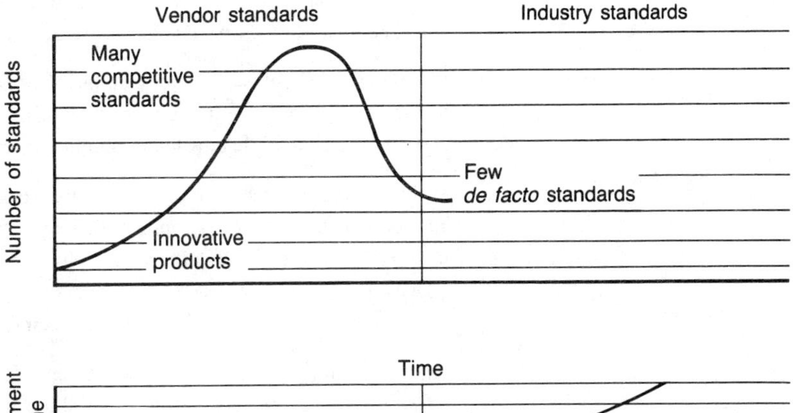

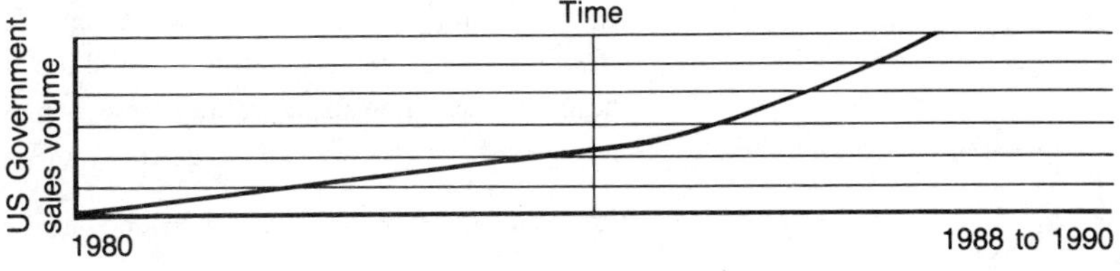

Figure 7.4 Emergence of industry standards (© X/Open Company Limited)

I would like to illustrate this with a couple of recent examples. This is what happened with the video cassette recorder market—a rapid growth in different standards between 1975 and 1981. In 1981 the VHS standard emerged out in front—and the market began to take off.

A second example, closer to home, happened in the personal computer industry. After the late seventies a variety of PC standards were introduced—the shake-out left IBM DOS machines and Apple systems out ahead. It was then, in the mid-eighties, when the market began to surge to the dizzy heights it has reached today.

Now look at UNIX—an open operating system and this is what happened. We have seen a convergence from 22 to 2 versions of UNIX, and the expected market growth is steepening into the nineties.

When X/Open started out in 1984 it set out to merge the alternatives into single platform standard accepted by both the market and the industry.

As late as 1987 our market researchers were forecasting a market size of around $20 billion for UNIX sales in the early 1990s. Now they are talking at between $35 and $40 billion—a doubling of the market. The frontrunners—AT&T's UNIX System V and the Open Software's OSF/1—are X/Open compliant developments and X/Open's platform XPG is closely aligned with the international standard, POSIX.

Now remember that this only refers to operating systems—and as already described there is much more to open systems than operating systems, and the total market potential is enormous.

Customer demands

The potential will grow with each step in the evolution of the customer's demands. As customer needs become more sophisticated, so the demand for compatibility among products becomes even greater. This has happened with telephones, televisions, audio systems, even cockpit instrumentation in the world's aircraft.

If we look at where we are today on the chart for open systems, we are only just at first base and we have just got there with the convergence of an open operating system. The next battle now being fought is over the human computer interface.

In the phase that we have now entered, the vision for the next year or so is one of a common open systems environment that encompasses data management, user interface, languages, and networking as well. That will allow the market in the early 1990s to develop a foundation for workstation systems—each with powerful networking capabilities and tailored to the specific needs of the individual users.

Add to the truly open system—that Open World of the year 2000—a successful information provider being able to combine the organization, the communication and the presentation of information into one system—that truly portable system—and you will see that we still have many more steps to take.

The open-system reality has only just begun.

Customers have the power

Now I want to give you a few ideas about the major factors reshaping our industry over the next few years. Many of these are happening all around us.

It is quite clear that one of the major differences between now and the early 1980s is the rising power of the customer. These days there is a lot to talk about the customer being King, but while the sentiment is correct, the reality turns out to be different because those customers still face a beggar's choice in many areas.

Yet, the customer experience curve is today as powerful a factor in the equation as the various technology learning curves. A much more knowledgeable, and much more powerful, customer base is shifting the industry's balance from the supply to the demand side of the industry.

Personalized computing

The customer is changing. More and more it is the line of business manager needing to arm his front-line technical and business professionals with 'the best for purpose' computing. The new industry of the next decade will be optimized to serve those professional end users who want to control the information that they have access to from their desktops.

It is such people—engineers, marketeers, and an increasing number of government professionals whose informed, or misinformed, judgements will make or break companies and organizations in this—the era that General Electric's Jack Welch has dubbed the 'white knuckled' decade—a decade of rising competition, of shortening product cycles and a decade of increasing diversity.

This will need a new structure that will form the basis of the new networked, multimedia industry specifically designed for the professional end-users who Peat Marwick, for instance, reckon already account for some 40 per cent of the total information systems spend.

It is an industry that will have to be based on open standards to allow user organizations to plan ahead—to create information networks that realize the potential of the information they carry—and to avoid the constraint of short-term product cycles and marketing strategies.

And those decisions are critical—the life-cycle of information far exceeds the life cycle of any systems and most applications. Think of any design or market analysis you need to do. In 10 to 20 years' time you may still need to be processing some of this same information that you gathered today. Open systems are going to help protect access to that information as the technologies, systems and applications change.

Value is the basis of open systems and of X/Open's mission. What that means is to look ahead, to work out with the users what the end-state might be and inject as much continuity and compatibility as we can into the planning process—that is the core of X/Open's business—value through compatibility and open systems.

There is only one way to find out what the international information systems community wants out of open systems and that is to ask them—so we did—over 400 professionals were asked from three continents during the last year.

The XTRA process

It was part of what we call the *XTRA* process—the X/Open process of bringing together all sectors of the industry to determine the next steps that are needed to get from where we are today to where we want to be tomorrow.

To be specific—XTRA 1989 produced an extensive list of over 100 key requirements that we have published in a document called *The Open Systems Directive*.

XTRA and the Open Systems Directive are the first leg of the journey to an Open 2000. We will be doing it again this year and we hope you will all support us.

What all this amounts to is that X/Open operates as a kind of independent international exchange—a true 'futures exchange' for the industry—where all the sectors can exchange and trade interests, ideas, expertise and needs to converge the best standards for the future, and it is where customer needs meet supply side interest to produce real, practical open systems solutions—it is where the rubber meets the road.

This is the way to transform those information ghettos into an open metropolis where value can be had, and useful answers given.

Call to action

All this needs action—action among systems suppliers, action among software companies, and action among users.

Users: you have a major increase in value at your fingertips. It's called Open Systems. It will provide you with choice—it will protect your investment—it will give you continuity of supply. So insist that your vendors put the plans for open computing into products. Specify X/Open compliance in all your future RFPs—and use X/Open to have your say in shaping the industry through the 1990s. Remember, you sign the checks—and you are the emerging great power—so use it and use it wisely in favour of open systems.

To conclude: X/Open is the change agent that can take the industry forward to an open 2000. Together suppliers and buyers can create the open systems opportunity. We can create a foundation of confidence in technology investment, of security in choice, and innovation for the future.

As the Bible says, 'Where there is no vision—the people perish'.

So, share our vision of an open future. Act to shape an open systems environment that will make an open future become a reality. Make this the open decade.

Geoffrey Morris, President and Chief Executive, X/Open Company Ltd.
Reproduced by kind permission of X/Open Ltd.

Summary: the role of X/Open

In this chapter, we have looked at the background, mission and work program of, potentially, one of the world's most powerful organizations—X/Open, which numbers among its members 20 of the world's largest computer companies.

We have seen that X/Open has been set up to define the technical requirements that will turn the theoretical concepts of open systems into a practical reality, and we have looked briefly into the means by which it is doing this. It is clear that the work has the support of very many individuals and organizations around the world, and that a genuine effort is being made to take account of the requirements of all sectors of the computer industry.

All this will be in vain, however, unless X/Open's own members commit to its stated objectives by:

1. **producing families of machines that support the X/Open CAE, and that are completely compatible both across their own, and across other members', product lines; and**
2. **marketing such systems aggressively and shipping them in volume.**

Without such obvious and visible support from its own members, the work of the X/Open company itself will be in vain.

If the users continue their market pull, and continue to make their requirements clear, as indeed they are increasingly doing, then those computer manufacturers that show their commitment by responding with products that truly meet the needs of open systems, will justifiably gather the reward.

CHAPTER 8

Users of open systems

In previous chapters, we recognized that many of the problems which are holding back efficient use of information technology are due to incompatibilities between systems. We saw that these problems would reduce dramatically if the computer industry would adopt the use of international standards, and suppliers would build products that conformed to them. Our basic premise was that this would enable the users of IT to concentrate on the real issues relevant to their associated businesses, and stop wasting valuable resources on re-invention.

In many cases, the issues appeared at first sight to be irrelevant to the users, because the costs that they create are often invisible. For example, users in one particular geographical territory do not often realize that the lack of standards in the handling of internationalization within software products causes additional costs in development and distribution that must be passed on to users in all territories.

We investigated the various segments of the computer industry market, and highlighted where many of the costs due to incompatibilities and lack of standards occur. This was followed by an analysis of the general standards that are needed in order that these unnecessary costs be reduced.

We found that there are two main areas requiring standardization. One of these is concerned with the ability to have software applications portable across machines from different manufacturers, while the other requires efficient communications to be provided between such machines. Combined, these two concepts would allow users to build distributed systems from whatever computers were appropriate, and use them to develop solutions required for some major new applications.

In order to understand standards developments, it was necessary to look at how standards are set in the computer industry. We overviewed the major groups involved in this work, and saw that this is not only a very difficult area technically, but also, because of the vested interests of the suppliers, it is difficult politically. We looked in detail at the activities of a particular organization, X/Open, which is supported by many of the most powerful participants of the industry, and which appears to be well-positioned to progress most of the issues, to the users' benefit.

In this chapter, we conclude by discussing the needs and concerns of the users themselves. By 'users' are generally meant the *purchasers* of the technology, among whom are many of the large suppliers themselves, of course. However, for our purposes here, the word 'user' will be taken to mean those commercial organizations, government agencies and educational establishments that are not themselves principally suppliers of IT products.

We will start by showing the mechanisms that currently exist to allow users to input to the process of defining the standards that are required. We will look at some groups that are trying to serve the needs of the users. This will be followed by a report on the formal process set up by X/Open for user input, with a summary of the results to date.

We will give some examples of users who have had the foresight and courage to move to open-systems procurement strategies early. We will show why they did so, and what some of the effects and problems have been.

Finally, we will conclude with some words of advice for users who wish to move to an open-systems procurement strategy, but perhaps will need some help in doing so.

User input to standardization processes

For the last few years, as standardization has become increasingly important for both users and suppliers of IT products, many organizations have struggled with the problem of how to get the 'real users' involved in the process.

In some respects, this could be thought an odd way for an industry to behave. In most commodity businesses, for example, the suppliers spend vast amounts of money trying to understand their users' desires, needs, and purchasing behaviour, often using very sophisticated techniques to do so, before developing any new product. In other words, they *start* by analysing the market.

It is only recently that the computer industry seems to have recognized the need to direct its efforts at solving the real problems of its users. For this reason, although one can describe the revolution that is going on at present as 'a move to standards', it can also be described as a complete turn of the computer industry from its previous, technologically driven mode to the emerging market-led stance.

Uniforum

The Uniforum group, previously known as /usr/group, is an organization set up in 1981 to serve the needs of users of products and services based on the use of UNIX and UNIX-like operating systems. Headquartered in the US, it works with many other affiliated groups worldwide.

Like many similar groups, the strategic directions for Uniforum are set by a board of directors elected by the general membership. These directors are usually volunteers who, while working full time for other organizations, are prepared to donate time and effort to the objectives of the Uniforum organization. Implementation of the strategy as set by the board is carried out by an executive team of full time employees of the group itself.

We saw earlier (Chapter 4) that Uniforum was responsible for the work which led to the first portable operating system standard, POSIX, and it continues to be very active in standardization work today. It is the leading group working on internationalization standards, and supports a number of committees working at the leading edge of technology.

To date, Uniforum has chosen not to seek accreditation as a formal standards-setting body, preferring instead to work closely with the IEEE and other groups. It has developed a process under which its pre-standards work is passed into the appropriate IEEE committee as consensus on the particular piece of work starts to emerge.

For many years, Uniforum, and groups like it, have tried hard to persuade their members to elect users to their board of directors. The desire has been to provide user input to discussions, the results of which could ultimately affect them dramatically. Until 1989 these efforts were in vain, but in that year, John Ozvath of MacDonalds Inc., an important UNIX user, was elected.

Apart from Mr Ozvath, the board of directors of Uniforum has been since the start made up of representatives from the large hardware manufacturers and the software suppliers. As we have pointed out, these are themselves usually large users of the technology, but they can only indirectly represent the interests of their own purchasers.

Uniforum is continually increasing its efforts to provide education on open systems to the marketplace, to encourage user input to standardization work and to promote the results. However, this work will be in vain if the results are not followed by the other participants in the process.

Corporation for Open Systems

In the US, the Corporation for Open Systems (COS) is an organization seeking to make OSI a commercial reality for users. COS was founded in 1985 as a non-profit organization, with members representing computer, communication, finance, manufacturing, and engineering industries, as well as agencies of the federal government, universities and other associations. Parallel organizations are the Standards Promotion and Application Group (SPAG) in Europe, and the Promotional Council for OSI (POSI) in Japan.

Although COS is vendor dominated, with only about a quarter of its voting members being users, major users among the members seem to feel that it provides a useful informal forum in which users and vendors can meet to discuss important issues, and where user input can be heard.

Other groups

There are many other groups which are similar in concept to Uniforum and COS operating around the world. Some of these specialize in portability issues, others in Interconnection, and some in particular components of the total applications environment.

Users can generally input to the standards process by participating in any or all of these groups, or by active participation in the formal standards bodies. Many large users dedicate huge amounts of resources to this in order to keep track of matters which are of strategic importance to them. For example, Visa International has a number of people tracking developments in communications and security, such issues being highly relevant to electronic funds transfer or credit checking and authorization, while Eastman-Kodak dedicates substantial resources to following standards work in image processing and graphics.

Such active participation is expensive. It has been estimated that the cost to an organization of having a person participate in a group such as Uniforum or COS can be of the order of $US 100 000 a year, while participation in a formal standards body can be more than $US 250 000. These estimates do not include the costs of the support given within the home company base, since these are not usually directly tracked.

Considering how much effort is dedicated by these companies to tracking the progress of standards, and how much possible duplication there is within it, common sense says that there must be a better way to deal with user input. This could be through organizations set up by the users, to define their own requirements. If sufficiently well supported, such a group could then impose those requirements on the computer suppliers through user procurement policies.

There are signs that such groups are in fact beginning to form. Within government advisory bodies especially, procurement policies often now require adherence to internationally approved standards. Commercial organizations are following the trend.

Information Technology Requirements Council

An example of such a user organization, the Information Technology Requirements Council (ITRC), was formed in 1989 by a group of powerful users in the US, including both government and industrial users. Such groups will probably emerge in other parts of the world in the future.

The ITRC was set up to be the primary body in the US for channelling user input to standards groups. Early members include the MAP/TOP User Group, as well as Aetna Life, Apple Computer, Bechtel, Boeing Computer Services, Eastman-Kodak, General Motors, Hewlett-Packard, and Xerox.

It is too early as yet to measure the effectiveness of this group, but clearly, with such a strong membership, it has the potential to become a powerful influencer of standards-setting priorities.

The X/Open Requirements Process

In 1989 X/Open instigated its Requirements Process, designed to identify the main concerns of IT users, and help it to prioritize its future open-systems standards work.

The first stage in the X/Open Requirements Process consisted of interviews with key people involved in defining technical strategy for over 30 major organizations in Europe, the US, and Japan. The group was made up of end users, independent software vendors, systems integrators, vendor associations and government agencies. The key output of these interviews was a set of requirements, arranged in a priority order.

The second stage in the process was the Requirements Conference, held in Montreal in July 1989. This brought together 104 senior representatives of organizations, from the same countries and industry sectors as the interview process, to discuss the issues identified in the first stage. The Conference ran for three days.

The third stage consisted of a questionnaire, based on the requirements identified in the interview process, to verify whether these requirements were in fact representative of the wider open-systems community. Over 430 respondents were surveyed in both Europe and the US, using telephone interviews.

The results of all three stages of the X/Open Requirements Process are contained in a report entitled *Open-Systems Initiative*, available from X/Open.

The X/Open Requirements Conference

At the X/Open Requirements Conference, working groups were formed from the participants to discuss specific subject areas, defined as a result of the interviews held at Stage 1. The working groups covered the following topics:

- open-systems directions;
- open-systems environment;
- security;
- system administration and management;
- human–computer interface;
- networking and communications;
- distributed applications;
- application development tools;
- data management;
- internationalization.

Each group was provided with background material that drew on the results of the interview process. Group members then generated, discussed and refined their requirements in that area, resulting in a structured requirement with title, abstract, description and business-oriented rationale.

At the end of the first day, each group presented their requirements to the conference as a whole, stimulating feedback from other groups with overlapping concerns. The working groups then refined and extended the set of requirements, resulting in a final presentation of requirements to the conference on the third day. At this time, participants assigned priorities to each of the total set of identified requirements.

The requirements most frequently rated as *absolutely essential* by the participants in the process were:

- standard human interface;
- global standards for networking and communications;
- applications interfaces for OSI networks;
- wider portfolio of open-systems products;
- distributed database access;
- transaction processing standards;
- products for migration.

There are no surprises in this list. All of these issues are acknowledged within the computer industry to be of great interest and concern to the users, and all of them have been discussed in various places throughout this book. In particular, it is well-known that a standard for the user interface is urgently required by many users.

Work within the computer industry on the human computer interface (HCI) is not focused on the definition of a single standard HCI. Instead, it centres on how to define an interface to it so as to allow at least two, and potentially more, of the products currently available from the large suppliers, to meet it. The developing concept is of a standard for an 'Applications programming interface' (API) which would enable an application developer to write a single program and have it able to use several user interfaces, and in particular, those supplied by Unix International (Open Look) and the Open Software Foundation (Motif).

The API approach may well solve some problems for the suppliers, and alleviate some for the applications developers, but it does little for the users. Their need is for a single interface that can be tailored to individual corporate needs, while not deviating from the basic standard. But the vested interests and heavy investments made already in this area by some suppliers make it difficult to give the users exactly what they want. In this respect, the situation for the user interface standard is no different from that for so much of the other standards work.

Conclusion on user input

Unless some way is found to put sufficient pressure on suppliers to eliminate the variability in standards, the initiatives for acquiring information on user priorities are in danger of being cosmetic only. In the end, the only real force is purchasing power, and this means users organizing themselves to take action, as we have seen them start to do.

Lessons from the fax industry

Before we move on to look at the world of open systems from the user viewpoint, we include here a recent editorial from one of the leading open systems periodicals, *UNIX World*, written by its highly regarded editor, David Flack. It provides a succinct summary of many of the points that we have tried to explain, albeit from a slightly different perspective.

> In these days of mass confusion in the computer industry—of standards struggles and desktop dilemmas—we could take a lesson from the fax machine business. Indulge me, if you will, in what is obviously an oversimplification to make a point.
>
> Like computers, fax machines were around for years before they became popular business tools. Also like computers, at first they were terribly expensive, terribly slow, and their quality and ease of use needed improvement.
>
> So how did fax machines become ubiquitous in just a few short years? And how did they become so popular so surreptitiously?
>
> Because fax machines are communications devices, manufacturers immediately understood that to work together, both literally and figuratively, they would have to agree on standards. They also understood that to improve slow performance, high cost, and poor image quality, they could not let standards get in the way of innova-

tion and improvement. Last, they understood that as fax machines improved, many people would have the older, slower machines and that the new machines would have to be compatible in order to communicate with these as well.

All during the time the fax industry was developing standards, I never heard a word about the IEEE or the ANSI committee meeting on fax standards. Nor do I recall the Canon people forming Fax International or the Ricoh people starting the Open Fax Foundation. All I remember is that when our office went to buy our first fax machine, the salesperson told us, 'No problem. They all work with each other.' And so far, no matter where in the world we have sent a fax, that's been the truth.

Unlike the fax industry, the computer industry has long relied on competition that places technical issues ahead of practical productivity concerns. Too often, computer vendors launch into a litany of technobabble based on meaningless benchmarks and absurd technological claims. Unable to assess the technical performance claims in relation to price, customers are often reconciled to guessing whether they made the right choice.

Fax machines are not a technological solution in search of a problem. They are technology applied to solving a problem. The reasons for buying a fax in the first place are also usually quite clear. If you have a six figure monthly Federal Express bill, you have a clear incentive to reduce costs.

There may be benchmarks for fax machines, but we have bought several here at *UNIX World* without any benchmarks and are very happy, thank you. When we upgraded, we understood the value and the costs involved. True, the choice was not very complicated, but there were many vendors to choose from, and when we finished the decision-making process, we felt good about our choice.

Now the fax industry could take some lessons from the computer industry. Here are my recommendations: Don't add features that prevent any other fax user from transmitting to, or receiving from, your latest machine. Don't design in special features that only work with other fax machines your company makes, no matter how much you think your customers want then. Don't add features just to keep the prices up in the face of declining costs. And for heaven's sake, keep the user interface simple. Don't make programming a fax machine so complicated that you need a four-week course just to send a letter to another business.

It would be a shame to mess up the fax industry just when the computer industry is beginning to catch on.

Editorial in *UNIX World* 1989, Editor-in-Chief David Flack.
Reproduced by kind permission of UNIX World Inc.

User case histories

Around the world, there are now some huge IT installations incorporating open-systems technologies. Although few of the users of these would say that implementation was easy, particularly those who made the move early, when many of the technologies and products were immature, none seems to have regretted their decision. Most are looking forward to the future, when they can expect to reap the long-term benefits of their strategy.

We give here a brief picture of the experiences of some of these, with thanks to them for allowing us to include material on their experiences.

Example 8.1 DHL: 'The world's largest international air express company'

Background

DHL is the world's largest international air express company, with a logo claimed to be seen in more countries than that of Coca Cola.

DHL ships more than 109 000 documents and parcels each and every day to and from more countries (172) than there are members of the United Nations. This involves more than 270 000 face-to-face customer contacts. An increasing number of shipments involve the use of computer technology.

DHL is heavily committed to the use of technology for efficient running of its business and to maintain its competitive edge, so much so that a statement to this effect is contained in its mission statement.

There are two main reasons behind DHL's commitment to technology. First, one of its founder owners, Larry Hilblom (the 'H' in DHL) has always insisted that technology play an important part in the business. For example, it is a requirement within DHL that all staff are capable of using a computer. Second, DHL views efficient use of technology as one of its best strategic weapons in an increasingly competitive market. For this reason, DHL concentrates on 'mission critical' systems—systems which must be justified by showing that they will improve customers' perception of how DHL does business.

The choice of open systems

In 1986, DHL made a decision to move to the use of open systems worldwide, away from many of its IBM System/36 computers. This decision was taken following an in-depth study, conducted by a team from within DHL, supplemented with outside consultants.

The choice of open systems, to be implemented with the use of the UNIX operating system, was driven by a need within DHL for portability of applications and people. The desire was to develop applications that could be run on machines from a variety of manufacturers and used in DHL offices around the world. If this could be achieved, then DHL personnel knowledgeable in the systems could carry that knowledge with them from site to site, as required.

Recognizing the increasing threat to paper and document delivery coming from EDI, Fax and electronic fund transfer (EFT), DHL also made the decision to focus on delivering objects, rather than paper. This anticipated a need for connections into clients, and other, computers.

For example, when goods are being carried to another country, the documents needed can be electronically transmitted ahead of time to the Customs and Excise authority, thus expediting the path of the goods through Customs. DHL expected that an open-systems strategy would ease these communications for the future.

Although using a proprietary international network today, DHL plans to move to an X.25 network to connect 155 DHL stations worldwide. Eventually, the system will comprise one giant network, capable of tracking and tracing a package anywhere in the world.

As yet, DHL has no EDI links operational, but is tracking the standards work in that area, with a view to implementing as soon as practical.

The hardware

The most important components of the DHL international network to date are fifteen large UNIX-based systems supplied by Pyramid Technology. These are running in the DHL headquarters in California, at its international gateway at Kennedy Airport, New York, and in centres in London, Virginia and Houston. In addition, there are more than forty NCR Tower UNIX systems, used in the UK, Africa and Latin America. These systems together, and other UNIX systems yet to be purchased, will be the critical components of the tracking and tracing system worldwide.

A key factor in the choice of manufacturer of hardware systems was the level of support and service offered in some of the regions in which DHL operates. For example, NCR was chosen as a supplier initially, not only because it supplied a suitable range of hardware, but because it was the only supplier offering a high level of service in certain areas of Africa. Having ported software to the NCR machines to supply those particular territories, other offices chose the same, or compatible, machines in order not to have to re-port the software.

All DHL airbills are bar-coded, and all couriers carry hand-held wand scanners. A packet is tracked by scanning it at all key points from sender to destination. Local conditions necessitated multiple, incompatible scanners in the overall systems. These all had to be integrated into the total system.

The problems

The learning curve involved in the move to open systems has been profound for DHL, a company which previously had been accustomed to the high level of support and service offered by a large manufacturer, IBM, for its proprietary technology, the System 36s. In the previous, mature environment, it had not been necessary for users within DHL to do other than concentrate on the business applications of the technology. Users experienced a severe culture shock when they moved to the immature UNIX world of the mid-1980s.

Although knowledge of the UNIX environment itself was very high within the suppliers, and much cooperative help was provided—for example, with the X.25 implementation, by Pyramid—nevertheless, DHL found that there were few people who understood both the environment DHL were going to and also that which they were coming from. DP professionals within DHL could have learned much faster if the bridges from the System 36 to UNIX systems had been defined, with differences and similarities pointed out.

Within the training programs available publicly, few trainers for UNIX systems were able to relate the new technology to the practical problems facing DHL in its business, concentrating instead on the internals of the system. In addition, where the System 36s are built to survive almost anything that an inexperienced operator may try to do, UNIX systems were not then, nor are they yet in general, supplied with such fail-safe mechanisms. DHL found by experience that an organization committing itself to open systems early must have, or have access to, strong technical support at all levels.

A commercial environment often requires machines to run for 24 hours a day, 356 days a year, at a cost effective price. Any housekeeping or maintenance work on the systems must usually be done while the machine is up and running. Although DHL is heavily reliant on its new system, it has not yet attempted to use fault-tolerant UNIX machines. Suitable configurations exist and have been evaluated however.

While the base of trained personnel with in-depth UNIX or OSI experience is increasing in the marketplace, such people are heavily in demand and are often more expensive than the people with proprietary know-how that might be currently employed. This can give personnel problems, when new people are brought in at higher salaries than long-term employees.

Some resistance can be expected from staff who have spent years investing in know-how for old, proprietary environments. Naturally, people do not want to throw their knowledge away easily. Again, products and services that help to bridge people, as well as products, from the old to the new would have been helpful to DHL.

Conclusions

DHL found that moving applications from one UNIX system to another was relatively easy. They also found it easy to link the UNIX machines, using the 3780 protocols, to the IBM minis and to the international network.

In general, DHL felt that the move to standards procurements would be easier if vendors understood the problems that the users face in managing the changes required, and if they provided more of the support services needed. In other words, the in-house MIS groups would prefer to concentrate on their own company's products and services, leaving the open-systems suppliers to provide working and workable solutions.

According to Mike Hurt, DHL MIS manager in the UK, 'We have vendor independence now. We need not choose to use it. In fact, it may be sensible not to do so. But if suppliers fall down on the job, it's good to know we can switch without penalty.'

Example 8.2 British Airways: 'the world's favourite airline'

British Airways is a giant organization, with commensurate ambitions. It would like to create an internal applications infrastructure that is independent of distance, culture and language.

It is a worldwide company with 10 000 of its employees working around the globe. And in each of 700 international offices in 90 countries, the local British Airways team has its own business to run, dealing regularly with local businesses and government.

The day-to-day administration of every office could, in theory, be run using a mixture of off-the-shelf business software. To put that theory into practice, British Airways plans to integrate local, commodity software into the company's global information architecture. The company's goals are specific: to keep training to an absolute minimum, and to develop common access procedures that are so simple that anyone in the company—from Tokyo to

Miami—can make immediate use of the local and remote applications available to them.

As one of the world's largest software houses, British Airways is acutely aware of the need to implement commodity software, so that its 1000-plus highly-prized specialists can concentrate on the airline-specific applications that enable British Airways both to use and market the best airline management systems in existence.

'Payroll, spreadsheets, word-processing, CAD/CAM are all products that we should be buying on the open market', says Bill Teather, Head of Tactical Management within British Airways' Information Management Department. 'So what we need are standards that allow us to integrate shelf products with home-grown systems. We want wider opportunities for collaborative systems development with other airlines, suppliers and software houses. And most important of all, we are looking for true internationalization, so that no matter where people work, what languages they speak, or what job they do, they will use systems to increase their effectiveness as members of British Airways' worldwide team'.

According to Teather, this means a commitment to international standards on an unprecedented scale. 'For us to develop a deep understanding of what the implications are and how we can help create these standards, we have to be part of the standards-making process,' he says.

To that end, British Airways became an early member of the X/Open User Advisory Council, a committee whose charter is to provide input to the establishment of X/Open's future technical and marketing priorities.

Mixing environments effectively

Central to British Airways' information architecture are the high-priority systems that deal with reservations, fares and departure control. 'These systems simply are not up for review,' says Teather. 'They already run to common IBM standards for the real-time systems that are critical to the success of any international airline. Neither do we plan to make any changes to our major business applications, such as aircraft maintenance—where we track the history of the one-quarter of a million parts that go to make up each aircraft in the fleet. Here, we are talking about many millions of lines of code that have taken many hundreds of man-years to create.

'We are looking for standards that respect that kind of investment and which help to provide quick and effective computer solutions at the grass roots level. We are looking for small, business management systems to run alongside, perhaps interact with, our mainstream airline management systems. And in particular, we are looking to introduce these systems quickly—with little or no training overhead,' notes Teather.

The hundreds of millions of pounds that British Airways has invested in IBM mainframes dwarf the thousands that have since been paid for some twenty or so UNIX-based systems, used primarily for small-scale modelling and decision-support applications. For the bulk of British Airway's work, UNIX simply is not yet suitable. To begin with, the heavyweight systems are developed largely by a consortium of airlines—all IBM users. And the high-profile, independent European projects for the development of the next generation of travel distribution systems, Galileo and Amadeus, have both standardized on IBM proprietary hardware.

British Airways is also a member of the X/Open User Advisory Council. Why does a firmly entrenched and well-directed organization like British Airways find it is so important to participate fully in the Council?

Making the X/Open connection

'Our motivation for joining is twofold,' explains Teather. 'First, we want to put across our views as to what we think standards should achieve. The idea is to make sure that real customers have a chance to put over the message—loud and clear—to those who develop the standards which we must all ultimately follow. And what is so refreshing about X/Open is that at last we have a standards body that puts users first, with a willingness and education to listen.

'Second, we want to learn more about the international standards scene, What technologies have to be standardized and integrated? How do standards bodies operate? Are the standards moving in the right direction? Which standards are important to us and which should we influence?'

In addition, the breadth of X/Open's involvement with creation of an open system architecture is particular appealing to a company like British Airways. The X/Open Common Application Environment (CAE) defines the interfaces for handling data management, applications integration, data communications, distributed systems, high-level languages, source code transfer and internationalization.

'A large number of important suppliers have already agreed to work together to create the X/Open common applications environment,' Teather continues. 'So it is clearly in our company's interests to monitor its development and, where indicated, put it into practice. In particular, we will be pressing very hard for a set of standards that not only helps us improve our own internationalization, but which will enable us to work more closely with the people with whom we frequently do business—aircraft suppliers, oil companies, freight carriers, hotel groups, travel chains, car rental companies—all of them on an international scale.'

For British Airways, the value of an open system environment extends to all facets of the company's operations. 'In 700 offices and airports around the world, we set out to provide a local service to the community that encourages people to fly with us,' explains Teather. 'Our staff speak their language, understand their culture, and follow their rules and regulations. It makes sense that our systems should do the same.'

Example 8.3 Shearson Lehman Hutton: 'trading in securities'

'Security' is usually regarded as one of the 'good' words. If you have *security*, you can borrow money. With job *security*, you can make long-term financial commitments. If you believe the software you are developing will have a broad base of hardware systems on which to execute, you feel *secure* in making the development investment. When you have a wide choice of compatible hardware and applications software within a particular information technology environment, you feel *secure* in asking for compliance to that environment.

The financial institution, Shearson Lehman Hutton (SLH), deals in securities of another kind. But the securities mentioned above are also critical to the solutions the company chooses to support its business priorities—and the vendors who will deliver them.

Solving the problems of a complex environment

Gary Handler has a challenging responsibility. As vice president for market decision systems at SLH, Handler is expected to be the master architect and general contractor for the complex of work stations and analytical programs that will become a strategic set of tools for his company's traders and salespeople.

'I need to create an environment for keeping accurate inventory of trader securities, for providing a real-time view of the market, for feeding pricing information into proprietary analytical programs, all to help determine whether a trade is a good one or not,' Handler explains.

For the company's salespeople, the decision systems should track the types of instruments their customer like, and help to define effective sales opportunities, linked to trader inventories.

'I am in the building stage now, but we do have many separate pieces of the solution residing on systems of different classes,' says Handler. His goal is to develop a platform for applications and analytical programs that is largely independent of the specific hardware upon which they function, and that would have the flexibility to take advantage of capabilities at all system levels.

Like others with similar responsibilities, Handler would much prefer to devote his resources to meeting his company's business needs and not have to divert any of them to working around connectivity and compatibility constraints.

The role of the CAE

'X/Open's CAE is critically important to my plans,' he states. 'I do not want to take a deterministic approach to this problem. Instead, I would rather let the problems drive the solutions.' Handler sees the common application environment (CAE) as providing the solution to those connectivity and compatibility discontinuities that today demand an inordinate degree of investment, with little or no return.

'I would like to be able to equip senior traders with sophisticated multitasking workstations, their assistants with less sophisticated workstations and perhaps others with dumb-terminal user systems. But I do not want to develop separate display applications for each type of user system. This is what is so appealing about a graphical interface like the CAE's X windows. It permits me easily to create compatible, yet screen-specific, graphics or text formats,' he explains. 'I can concentrate on the users' needs without having to worry about compatibility issues.'

Handler is pleased that the CAE is extending beyond its mid-size system roots. Recent conformant product announcements from small-system and large-system software vendors is an encouraging trend to someone whose solutions are likely to be hosted in a hierarchical system, involving everything from mainframes to desktop systems.

In search of the ultimate system

Handler's ultimate system would provide a combination of product-specific view and capital markets view, permitting the company to determine its overall exposure, and to determine the 'weighting' of trades, one to another. He has identified some strong commonalities in applications from product area to product area, and wants to build a 'heterogeneous network' with a tremendous degree of 'data commonality'.

In the long run, as the CAE succeeds in gathering a critical mass of users, and hardware- and software-vendors around it, it may create a 'consensus of format for market-supplied data', Handler says hopefully. But short of that, if it just created an environment where 'my spreadsheet modules understand and can process the data exactly as they are organized and formatted in my databases', Handler believes it would be a big step forward.

From Handler's perspective, the prime value of X/Open's CAE is not in any of its specifications but, rather, in its ability to rally a critical mass of vendors around it. 'The real bottom line for me is the range of applications, tools and hardware platforms developed to comply with the CAE,' he offers. 'I have two excellent vendors now, who attack the underlying problems from two very different angles. One takes a communications approach; the other takes a data management approach. The extent that these tools are compatible is pure accident. But if both were developed to meet the specifications of the CAE, compatibility would be given. The CAE could provide that missing glue.'

Migration experiences

The biggest issue facing a company looking to move to open systems is not usually whether the decision itself is the correct one. Once the basic factors are understood, the logic of the approach is inescapable, and the potential benefits obvious. The real problems are those of migration, integration and management of the change process, i.e. 'Given that we want to get to Open Systems within our organization, how do we do it from our present position, while continuing to run the business efficiently?'

In the following, we give two accounts of installations that have struggled with this issue, one relating to installation of UNIX systems and the other to OSI, specifically an X.400 electronic mail application. Although both experienced problems, these had little to do with the specific open-systems technologies adopted. Most are common to any project that involves making major and complex changes to a real world system.

Migration example 8.4 Introducing UNIX into non-UNIX environments

The following is extracted from a paper presented at Uniforum 1990 by David M. Sherr of Shearson Lehman Hutton, Inc. and appears here with his kind permission:

The decision to integrate new technologies into a business organization is complex. There are conceptual and political aspects to consider, as well as the more obvious technical components. The problem of installing UNIX systems is not unique in this regard.

The management decision to acquire UNIX systems has to be driven by the business issues, and not by the technology. Introducing UNIX can be facilitated by:

- the existence of specific vertical software;
- the commodity pricing of hardware running UNIX;
- the emergence of C as a standard language;
- the growing UNIX-trained workforce;
- the ascension of UNIX advocates into decision-making and decision-influencing positions.

Factors mitigating against UNIX include:

- the capital invested in existing proprietary software and hardware;
- the relatively low cost of enhancing and augmenting these existing environments;
- the corps of managers and technicians whose careers and livelihoods are dependent upon the continued maintenance of the non-UNIX systems.

Beyond CAD/CAM, the financial services industry has been a most fertile environment for UNIX. UNIX-based systems tend to be more flexible than information processors running proprietary operating systems. Flexibility is the key to the rapid response needed to the ever-changing conditions in the financial marketplace.

The disorder caused by the last eight years of financial deregulation has made all information processing systems inadequate at a very rapid rate. This has opened management to thinking of solutions in new ways, especially if these are flexible and keep options open.

Control of the treasury and capital of the corporation will remain on proprietary systems for the foreseeable future. It is one thing to make money using UNIX-based trading and market information systems, and it is yet another to keep the wealth created.

Three UNIX-based introductions into non-UNIX environments supply most of the data for my comments. The three UNIX-based systems were (1) a portfolio management and accounting system for an insurance holding company, (2) a floor display system for a regional stock exchange and (3) a fixed income on-line deal capture system for the investment bank trading floor of a major international bank.

The portfolio system involved development over a four year period, and now runs on an AT&T UNIX system, with seven users. The non-UNIX environment into which it was inserted consisted of a time-shared UNIVAC mainframe system. The main data processing group dealt with Honeywell systems and was converting to an IBM mainframe.

The new system runs in the investment department and is continually being enhanced by both users and developers. Since the system's installation, the annual transactions have increased by about 135 per cent and several new financial instruments have been added, along with many new reports. This system has now operated successfully for two years.

The investment departments of insurance companies have needs which are radically different from their policy/underwriting groups. Their investment guidelines tend to be more conservative than their underwriting risk-taking. Policy transactions run in the millions each year, while the investment department handles several thousand. In this particular case, the investment department was run by a fiercely independent group of managers, who fought to remain autonomous.

In retrospect, 18 months could have been cut off the project if the old system had first been replicated on the UNIX platform. Instead, a 4GL development platform was built to support the programming of the new system, and there was reluctance to use the new system until it had been fully proven. The resources were not available to support both systems concurrently. Thus, the development team had to spend time processing the transactions *and* keeping the two systems in synchronization. Had the old system been programmed on the UNIX platform first, this would have made the new system the only variable changing at any one time.

Through all the problems with implementing the portfolio system—and they were many—the clients were patient. Today however, the system has no sooner been established than the DP/MIS mainframers are eyeing it for replacement.

The stock exchange floor system runs on duplexed AT&T UNIX systems, driving two ethernet networks. To these are connected 10 IBM PC compatible systems, which in turn drive about 280 overhead floor monitors.

The system in place at the time of the transition was a 10-year old DEC minicomputer, which was driving the floor monitors, and was itself controlled by a Honeywell mainframe computer. This system was difficult to operate, arduous to maintain and impossible to enhance. Additionally, the proposed new architecture fit the business plans of the client—a need for new products, faster response to competition and improved services to members. The new system, which was 'state-of-the-art' at the time, involving both UNIX and Ethernet, has so far assumed control of approximately half the floor monitors.

This system is a highly specialized application with distinct parts which could be built, tested and demonstrated incrementally. Additionally, since the processing was distributed, the trading floor could be brought on-line in parts, for the options, futures and indexes trading pits. Serial network managers were developed and used first, replaced by the ethernet managers at the final stage.

The investment bank trading floor deal capture system has so far taken one year for development of a prototype and a pilot test. The prototype system was developed on networked SUN Microsystems systems. Back office processing, to which this system interfaces, again runs in an IBM mainframe environment.

The eventual system will be integrated into a total trading floor system with more than 325 trading desks. The system provides internal and external rates, market information, access to sales support tools, and advanced proprietary models of money and markets.

A common feature of all three of these systems was the requirement that each interface with non-UNIX systems. The ability to co-exist comfortably with, and indeed to promote, the use of heterogeneous environments is a great strength of UNIX. To varying degrees, each of the systems is now able to be grown or contracted, depending on the business situation.

To any organization wishing to install UNIX systems into a non-UNIX environment, I offer this advice, based on my own experiences:

- Pick defensible applications well suited to UNIX.
- Use an incremental implementation, if at all possible.
- Present no surprises to the other, proprietary, systems that are installed.

Finally, a quote seems to sum up the main limiting factors to introducing UNIX into non-UNIX environments:

'There is nothing more difficult to plan, more doubtful of success, nor more dangerous to manage than the creation of a new system. For the initiator has the enmity of all who would profit from the preservation of the old system and merely lukewarm defenders in those who would gain from the new one.'

This was written in 1513 by Machiavelli for Lorenzo the Magnificent of the Medici's, the powerful banking family which ruled Florence for almost 500 years. Lorenzo thanked him and said he would take it under advisement. Machiavelli was never a political force in Florence after that, although history reveres him for his political insights.

Is there a message for us here?

Migration example 8.5 An OSI application case study

This example is taken from the UK Department of Trade and Industry (DTI) Information Engineering Directorate (IED), and comprises a description of the installation of an X.400 Electronic Mail System. It is included here with the kind permission of the UK DTI.

Introduction

The Information Engineering Directorate (IED) of the UK Department of Trade and Industry (DTI) is responsible for funding research and development into information technology cooperatively between UK industry, research institutions and the academic community. Its purpose is to ensure that maximum advantage is taken by UK industry of advances in technology. This example describes the establishment of an X.400 based electronic message handling system within IED and highlights some of the implementation problems.

Objectives of the IED X.400 system

The IED needs to maintain effective communication with its partners in the industrial community, the academic community and other government departments.

There is a continuing requirement for the exchange of proposals, reports and a host of general correspondence by IED and the organizations it supports. Most of the text originates on electronic office equipment and it is natural that it should be mailed through electronic mail. Until December 1988, communication was handled by a proprietary X.25 dial-up mail server, Alvey Mail, on which all the communicating partners were registered.

The objectives of the replacement electronic mail system were therefore:

- to integrate the partners in the industrial and academic communities and those in the government departments into a single electronic mail network;
- to provide a general improvement in facilities over the Alvey Mail system;
- to provide an open-systems solution based on international standards;
- to provide a replacement system which could be integrated with the electronic mail systems already used by many of the correspondents.

Based on these objectives, it was decided to replace the existing system by one conforming to X.400 standards and link a number of different existing X.400 services together to provide good electronic mail communications as required within the community of users involved with the IED.

Background

The IED had already invested in proprietary office automation systems from Xionics for managing the administration of the research programme. This system, installed in 1983, provided facilities for capturing documents, editing them and transferring them to other users of the system. There were 100 terminals attached to this system and its associated printing equipment, and there was a link to the Alvey Mail system.

The IED wanted a solution which could more effectively integrate the mailing systems of the partners, with less of the financial burden for the service being placed on the IED itself. The choice of system had, of course, to integrate with the existing IED office and mailing system already in place.

A series of investigations were carried out in early 1988, following which the decision was taken to purchase a gateway system that would provide a link from the proprietary Xionics office system in IED to the X.400 world.

At this time, all the academic partners were linked into the UK's Joint Academic Network (JANET), which was about to have its own X.400 gateway. Also, there was already a mailing system for government departments in use, the Inter-Departmental Electronic Mail (IDEM) system, also based on X.400, running as a pilot system. Furthermore, British Telecom had established its GOLD 400 service in early 1988, which not only allows the integration of private X.400 systems, but allows users of the Telecom Gold mail service to connect to X.400 via a gateway.

It was decided that all three of these X.400 systems—JANET, IDEM and GOLD 400—should be integrated to form a network which would satisfy the needs of all the IED partners. The plan was to do this by the end of 1988.

Decision timetable

In February 1988, the timetable for the project was agreed as follows:

May — Install the X.400 gateway. This was to be supplied by Xionics and was based on an INTEL 386-based PC running UNIX System V, with portable software from Retix Inc. providing the X.400 capability.

June	Connect the gateway to BT's GOLD 400 for engineering activity and trials. This was to ensure that both the private X.400 system connection through GOLD 400 and the link to Telecom Gold worked in a satisfactory way.
July	Test traffic through to the IDEM and JANET gateways. Produce a feasibility report discussing how the interconnection of the systems will be achieved.
Aug.	First trial of the systems with JANET participation.
Nov.	User tests with Telecom Gold subscribers. Between October and November, transfer the operational traffic from the old to the new system. Retire the old system.

Network topology

The proposed IED X.400 network (Figure 8.1) depended upon the availability and use of the GOLD 400 service from British Telecom. It was anticipated that there would be over 100 users from the DTI IED using the network, along with a similar number from other areas of government on the IDEM system, plus approximately 300 Telecom Gold users and 300 academic users of JANET. The network made use of an X.25 service between the partners.

It was expected that private industries (through Industry Private Management Domains, IPMDs) would also connect into the GOLD 400 network and thus also indirectly into the IED network, and that industry partners might also connect directly to the IED network.

Implementation

Early activity

A series of reviews were established to manage the implementation process. These occurred every two to three weeks and involved representatives from each of the participating areas.

At the first of these meetings, an activity to detail the capabilities of each of the contributing X.400 systems was started in order to establish a set of common factors. To this end, a questionnaire was prepared and sent to each of the groups, and an analysis of the replies made in early July 1988.

Following a meeting with British Telecom in May 1988, it was decided that:

- A group registration could be established within GOLD 400 under the auspices of the IED. This meant that individuals would not need to register separately on the GOLD 400 service.
- Billing would be at group level to a minimum charge for the group.

An issue was raised about how the GOLD 400 directories were to be updated. These form a key link between the Telecom Gold users and those accessing the network via GOLD 400.

Results from the survey

At a meeting in early July 1988, the questionnaires were reviewed, with the result that a number of incompatibilities were identified between the systems to be integrated. There were four major issues identified:

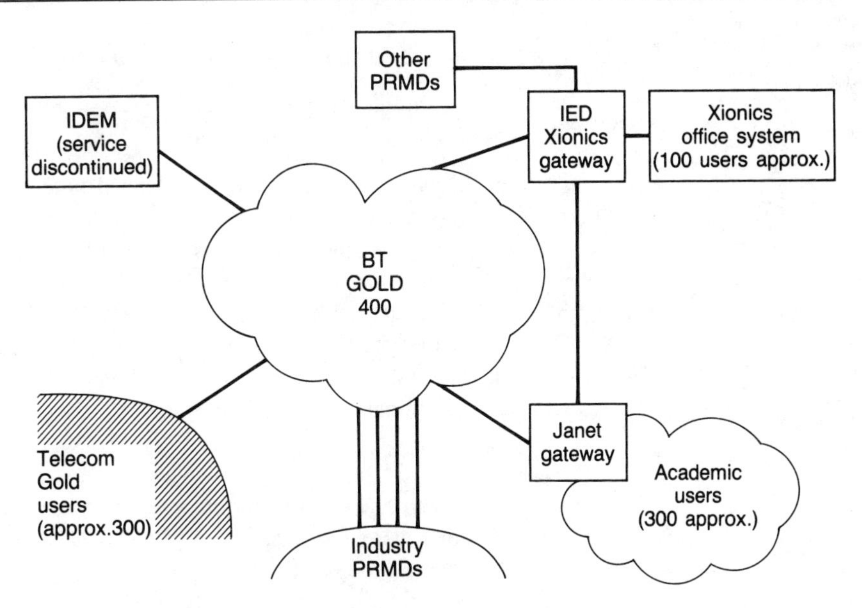

Figure 8.1 IED mail network topology

The first issue involved ensuring that forwarded mail items and reply requests were handled correctly and consistently between the systems.

The second meant ensuring that users of the system understood that they might receive delivery reports from gateways, which may not necessarily mean that the mail item had been delivered to the expected recipient.

The third problem related to commercial considerations of the GOLD 400 service. First, any user from a private system who wished to reach, or be reached by, a Telecom Gold user, needed to be included in the Telecom Gold directory. This meant that all potential users of the IED mail network had to be registered on the GOLD 400 directory, which could prove costly to maintain.

Furthermore, British Telecom were reluctant to accept JANET as an organization, because this contravened their own definitions. For example, JANET could accept a group of users under the organization name 'Manchester', accepting that it stood for Manchester University. This was unacceptable to British Telecom, because it must ensure that the X.400 mail system addressing meets the needs of the whole of UK industry, where 'Manchester' would mean the city itself.

The fourth issue related to 'Deferred Delivery'. The BT GOLD 400 charging mechanisms gave substantial savings for mail sent outside prime time. Unfortunately, none of the X.400 gateway systems had implemented Deferred Delivery and so mail would be sent whenever available, rather than when it was most economical or appropriate for it to be sent. The underlying Retix software used in the X.400 Xionics Gateway could support message priorities, however, and be configured to send messages at appropriate or economical times.

Xionics X.400 gateway demonstration

In the middle of June, Xionics demonstrated their X.400 gateway capability. A number of issues came up during this:

- Much expert guidance was needed before the X.25 PAD and the X.400 systems were configured properly.
- Very few diagnostics were available to the system manager to indicate when errors had occurred. Moreover, there were no management statistics associated with the mail software or the Message Transfer Agent software. Without these, it would be impossible to predict or report on trends.
- The directory management software needed enhancements to both its functionality and its user interface.

The revised schedule

Following the demonstration, a review was held of all the outstanding issues. Actions were assigned either to clear them or accept them as being outside the specification of the system.

A timetable was drawn up for adding X.400 capability to the IED office system. This identified that, given an order placement date of the middle of August, the Xionics Message Transfer Agent (MTA) and its software would not be delivered until September. This would allow most of October for trials of interworking between domains, training the system administrators, and the creation and evaluation of user documentation.

During November it was expected to train the users, issue the documentation to them, and have the gateway fully operational. The existing mail system was expected to be retired by the end of December.

Interworking between Telecom Gold and JANET

While investigating the problems of addressing, a serious difference was found between Telecom Gold and JANET regarding the use of names.

Telecom Gold assumed that commercial electronic mail users would all belong to an organization. It therefore insisted that everyone who received mail sent from a Telecom Gold user interface be described in terms of his or her name *and* organization. The GOLD 400 directory then provided a look up for the name for routing the message from the unique organization name and organizational units.

JANET, on the other hand, assumed that within a private system, such as the UK academic community, free use can be made of all names, including those of organizations. JANET therefore defined an organization-naming structure which is unique within JANET but at odds with the BT interpretation.

The possibility was discussed of programming the JANET X.400 gateway in order to ensure that messages within the JANET community could be successfully delivered.

The other issue raised by this interworking difficulty was a commercial one. British Telecom charge for each organization entry in its GOLD 400 directory and hence would wish to levy a significant charge for registering *all* the organizations of the JANET network on the GOLD 400 network directly. From JANET's point of view, this was out of the question.

Installation

The Xionics MTA was finally delivered and installed in late November. During December, the database administrators were trained on the system management of both the Xionics office automation system and the Xionics MTA. An important part of this training was the establishment of routing tables to access others on the network.

An independent decision was taken at about this time to close down the IDEM service.

Situation as of April 1989

Apart from the connections to Telecom Gold via GOLD 400 and to JANET, there are six other Private Management Domains included in the Xionics system routing tables. These were requested by staff in the IED who needed to communicate with these groups.

Of the 110 users of the Xionics office automation system in the IED, at least 25 are using the X.400 gateway, sending between 50-60 X.400 messages per week, mostly to the JANET gateway.

All users registered on the Xionics system are included in the GOLD 400 directory. There are currently between 30 and 35 requests outstanding for mailboxes on the GOLD service, which have not yet been allocated because of the addressing issues between JANET and the GOLD 400 directory.

The Alvey Mail system was eventually closed down in February 1989, although all funding from the IED ceased as of December 1988.

In order to help with the management of directories and provide a directory facility to users, a simple off-line directory was produced. This is updated monthly and distributed to all participants in the network.

Benefits gained

The benefits gained by implementing this X.400 system within IED are described as follows:

- Each of the participants in the network now has the ability to support their partnership communications through a single service.
- The X.400 capability of the Xionics gateway allows the IED to access a far wider audience for mail than was possible using the old system. For example, there are already connections into other European networks.
- The IED no longer carries the cost of managing a centralized mail server within its administration budgets. The onus is now firmly on the message originator to pay the cost of his own communications.

Some problems remain:

- Because of the addressing difficulties, IED have not yet succeeded in establishing connection from a number of the industrial users in Telecom

Gold to the JANET academic network. However, as these industrial users themselves become X.400 PRMDs they should be able to access the JANET network by connecting to GOLD 400.

- The difficulty of registering partners in the academic community on GOLD 400 persists.

Key learning points

The key points to be learned from this example are as follows:

- It is vitally important to ensure that there is a standard for naming and addressing on networks. For system implementors, it is important to ensure that manufacturers conform to these.
- Providers of network products need to be aware of the implications of addressing constraints.
- There is a need for networking skills at the transport layer, where a number of problems occur in mail services of this kind. The requirement is for such things as configuration skills, debugging software and line analysers.
- The software in many X.400 mail products is generally poor in such areas as the user interface, the system manager interface, and the network manager interface.
- Network management statistical and debugging tools are not commonly available. Network management systems to help solve the problems which the database administrator faces will probably not be available for some years and there is likely to be a high cost paid for network management systems in terms of system resources used.
- In order to integrate networks, there is an urgent need for a single directory service. Each of the systems on this network came with its own directory look-up systems but there was no common facility.

Until internationally standard X.500 directory systems become commonly available, which is not expected until 1991, there will be a need to maintain proprietary directory systems—perhaps even several for one network, probably with much overlapping of entries.

Advice for users of open systems

Before committing to an open-systems strategy, users of Information Technology should require satisfactory answers to a number of questions. Amongst these are the following:

1. Why adopt open systems?

- What are its advantages over older technologies for data communication? for applications support?
- Are open systems the best solution for my particular problem?
- Will open systems make life easier?
- Will open systems save money? or resources?
- When will the pay-off be seen?

2. What constitutes an open system?

- What manufacturers' products are 'open'?
- How can conformance be checked?

- What components are required to make an open system? An open-systems network?
- Does the intended application mix affect the components required?

3. *Is it practical to adopt open systems?*

- Are sufficient conforming hardware and software products, and data transport services available to make the implementation of open-systems techniques a practical reality?
- Do they work?
- Are the support services available?
- Can the organization handle the necessary changes?
- Can the security of data be assured in the context of open systems? How?
- How does the security of open systems compare with that of other systems?
- What about the administration of an open network?

4. *What about migration?*

- How efficient are open-systems techniques in comparison with established communications methods for both local and wide-area networking?
- How easily can standard POSIX and OSI systems co-exist with established operating systems and communications methods?
- Is migration from established systems and methods to open systems a practical proposition?
- What products and services are available to help in the migration and integration process?

5. *What other considerations are involved?*

- What external pressures—from common carriers, government, customers, suppliers, competitors, employees and others act in favor of a complete or partial commitment to open systems?
- How soon will this pressure be felt?

We hope that with the help of the material covered in this book, updated over time as technology changes, users will be able to put at least some of these questions into a relevant context and even, to some extent, answer them. Addresses of organizations that may supply further information and help are listed on page 254.

Now that the open-systems market has achieved critical mass, many companies are emerging to help users with implementation. In the early days, most of these will be systems integrators, with knowledge of certain proprietary environments to which the new technologies will be added.

Manufacturers themselves are gearing up to offer more help in this area, recognizing that their ability to co-exist with machines from other suppliers will be a major key to their future survival.

In all, there are many sources available for various levels of advice and support. For many users, the problem is not going to be whether or not to implement an open-systems strategy. The problem is, rather, how to implement it, given the base of systems from which the organization must migrate, and what help is available in the management of the change process. The problems of migration and integration must not be underestimated.

Conclusion

For the first time in its history, the direction of the computer industry is to be controlled by its users—the purchasers of its technologies. For many years, these users have been frustrated by the seeming inability of their suppliers to respond to their needs.

Now the users must take the power that is offered to them, and use it to help the technologists to put their brilliant innovations to practical use. Unimpeded by the vested interests which prevent the computer manufacturers from taking those decisions which would maximally benefit the market as a whole, the users must assume this responsibility themselves. Only then will the potential benefits of 'The Information Age' be made easily accessible for all of us.

Appendix 1 Data handling

The most common reason for organizations to buy computers is in order to process and to store data. They invest in communications equipment to move data from one place to another. They train staff to enter and to interpret data. It is not surprising, therefore, that many of the standards for computing and communications are concerned with the representation, structure, and manipulation of data.

The user view

The biggest problem with data standards is that, although there are many of them, both *de facto* and public, most of these address low-level issues, rather than the high-level problems faced in real life.

For example, at the very lowest level ISO lists 45 broadly compatible standards related to the representation of the characters which make up text within computers. But none of these standards relate to the idiographic or syllabic texts used widely in Asia. So work is now in progress (on Draft Proposal 10646) to enable representation of all the world's alphabets.

To compound this very basic problem of data representation, many countries have standard national variants of the earliest international standard in the area (ISO 646:1983), while many suppliers of computer equipment diverge from those standards. For example, the Apple Macintosh and the IBM PC differ from each other, and from international standards, in the codes that they use to represent accented characters (such as à, é, î, õ, and ü) and while the PC goes beyond any public standard by having a box-drawing character set, the Macintosh does not.

Perhaps the most widespread example of non-standard character representation exists in the IBM-compatible mainframe world, which uses Extended Binary Coded Decimal Interchange Code (EBCDIC). Almost all other types of computer use variants of the American Standard Code for Information Interchange (ASCII) and of the other public standards which have grown from it.

The fact that many millions of users communicate with mainframe computers by using PCs, unaware that the two types of machine represent data in different ways, suggests that translation at this level can be a trivial task. However, the majority of computer information is not just characters. Data in a database is usually held in some codified manner. Accessing a particular data item, and making sense of it once it has been translated, is a different matter.

Different classes of computer use a wide variety of different database management systems to record and retrieve data. The result is that the computers in a large organization can become 'islands of information', each using a different data management scheme and hence different methods of internal representation (see Figure A1.1).

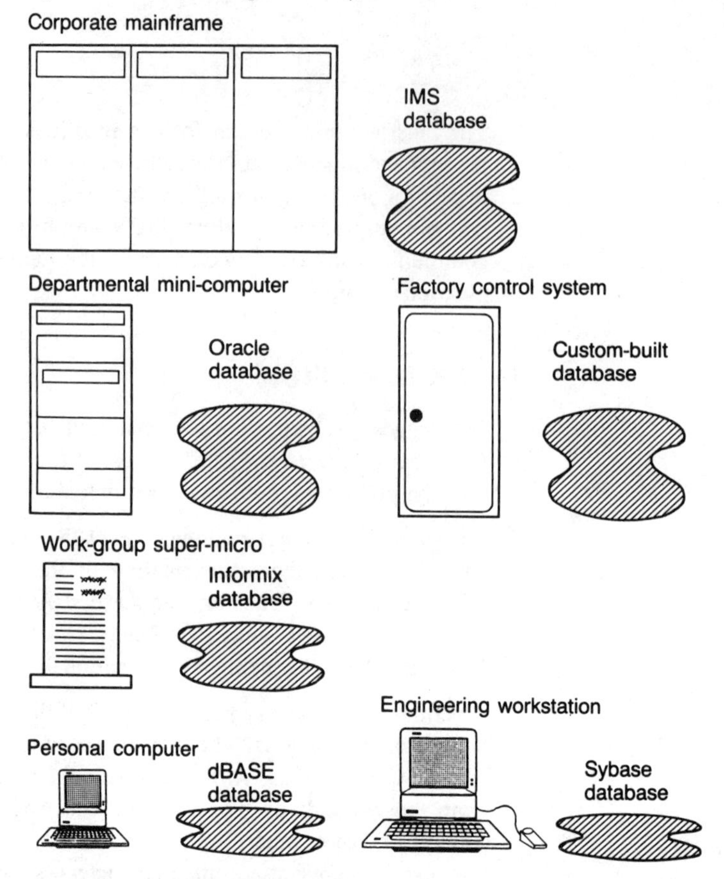

Figure A1.1 Islands of information

These islands can be connected using a variety of communications techniques, as shown in Figure A1.2. But this does not solve the basic problem. In order for one system to be able to access and interpret a particular class of information held on another, a special-purpose program must be written to retrieve the data and to convert it into a form which the destination system can handle.

Although an improvement on isolated islands of information, the connected islands scheme requires that many special-purpose programs be written in order to answer specific questions about the corporate information base. It is then likely to be difficult to get fast answers to questions, particularly given the long lead-times required for the production of new computer programs in most organizations. In addition, there will be a considerable overhead required to maintain such systems.

The need for specialized knowledge about particular data management and communications systems, and the need for special-purpose programming

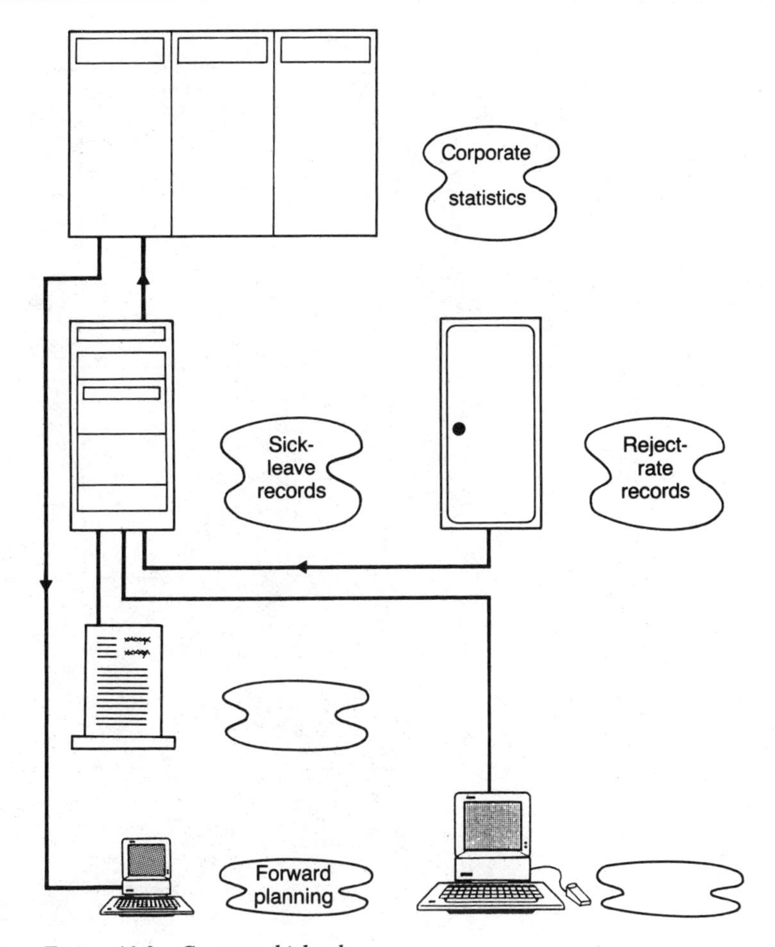

Figure A1.2 Connected islands

could be radically reduced if the computers of an organization could be made to appear as a consistent whole, with complete connectivity between all systems, and with simple, uniform access to all information, wherever held.

The word 'appear' is important. Provided that the computer installation appears consistent to its users, it does not matter to them how the effect is achieved. It may be that it uses several different types of database manager, and that there are not direct connections from every computer to every other. Provided these differences are hidden, the users need neither know nor care about the implementation. However, the differences may be extremely important to the systems manager.

Standards for data handling are moving slowly towards providing uniform access to data, wherever it is held (Figure A1.3). Of course, such widespread connectivity brings its own perils, particularly in the area of security. Unauthorized users must be denied access to sensitive data, and prevented from tampering with critical computer systems and communications links.

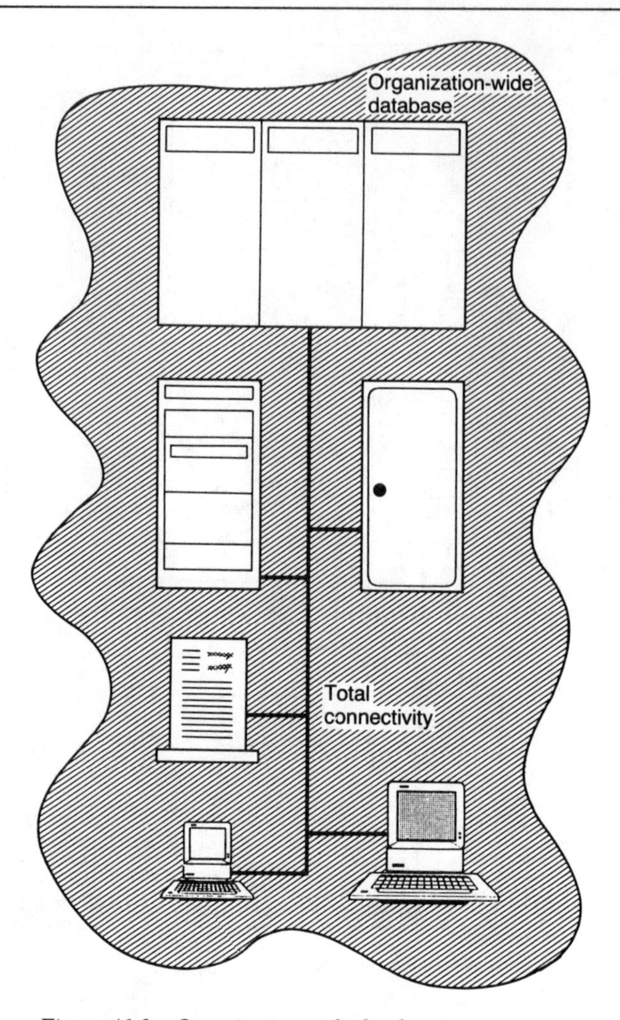

Figure A1.3 Organization-wide database

Proprietary database management products, such as Oracle and Ingres, not only keep up with standards but are able to move ahead of them. Implemented on systems from many different suppliers, these and other database products allow the sharing and exchange of data across systems. This is achieved by using the same product on all systems, or by using an alternative scheme for which the database supplier has provided an interface.

Of course, by purchasing a single database management system that supports a wide range of computer types, an organization risks becoming 'locked in' to a single supplier's software product. The hardware lock-ins of the 1970s are then replaced with software lock-ins in the 1990s. Clearly, this defeats the purpose of open systems, which requires freedom of customer choice, both in hardware and in software.

Leading suppliers of database management software now realize that they must supply interfaces to other suppliers' systems because enormous volumes of information are held in formats which are not, and never will be, standardized. IBM's mainframe product, Information Management System (IMS) is a case in point. Although DB2, a newer IBM product, is recommended for new applications, the amount of data held by, and applications written for, IMS is so vast that most of its users are not likely to convert to DB2's newer, and more open, format.

Even though IBM provides migration tools to aid the conversion, the cost and inconvenience of the process is often too great for the users to accept. This is a general problem applicable to any major change. Even when standard, uniform schemes for data management become prevalent, it will be necessary to provide access to data held in alternative, older formats.

Figure A1.4 suggests that data handling occupies only a small part of the system software picture.

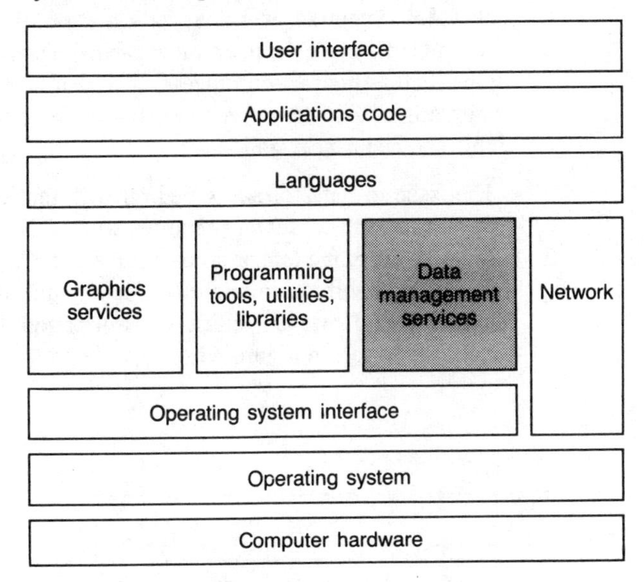

Figure A1.4 Apparent scope of standards for data handling

In practice, standards in almost every other area of information processing affect data handling:

- The user interface captures and displays data.
- The applications code manipulates data.
- Programming languages provide the means of manipulating data.
- Graphics services are often driven by complex data structures.
- Utilities and libraries ease the handling of data and provide additional functionality, such as enhanced security.
- Networks transport data.
- Data passes back and forth through the operating system interface.
- The underlying hardware stores and retrieves data.

Only the operating system itself is relatively independent of the content and structure of the data that it passes between the hardware and higher levels.

User needs

Broadly, users want data to appear in the right place at the right time, and in the right form. Data which appears in the wrong place is a security risk. Data appearing at the wrong time is usually too late. Data in the wrong form is difficult or impossible to understand and use.

Security is an issue which affects every part of an organization and its data processing.

Timeliness is a function of two variables: the amount of power available to perform a function, and the amount of work required by the function. Thus a powerful mainframe computer should be able to perform a complex task more quickly than a personal computer. Equally important is that of how quickly the available power can be harnessed to perform a function. Little is gained if the mainframe that can complete a particular job in seconds cannot schedule that job for several hours because of its outstanding workload. A personal computer working full-time on the same job may be able to complete it before the mainframe can even start on it. On a shorter time-scale, jobs such as transaction-processing or real-time control must be completed in fractions of a second. Even the shortest wait for resources to become available can be unacceptable.

The issue of data format is tied up with that of timeliness. Translating data from one form to another requires work. The more work that is required to perform a job, the longer it takes (unless the power applied is increased). It is desirable, therefore, to minimize the amount of translating needed for a particular job. However, much transformation is unavoidable. Consider the example shown in Figure A1.5:

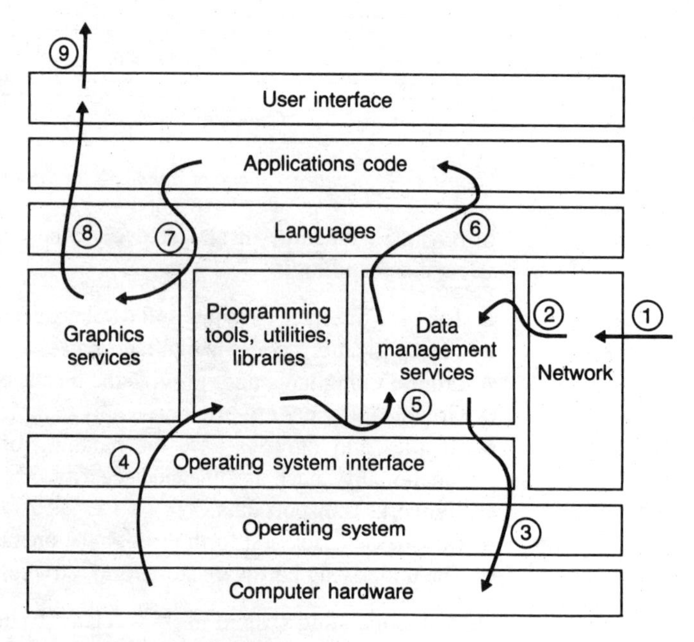

Figure A1.5 Data handling example

1. Data arrives on a computer system over a network. The received messages contain addressing information and codes allowing the correction and detection of errors, which must be processed and removed before the data can be stored.
2. The data passes to a data management system, which augments it with indexing information so that it can be retrieved rapidly when needed, and logging information, so that the contents of the storage system can be reconstructed in the event of a failure.
3. From here, the information passes through the operating system interface and the operating system itself to the hardware. As a result, information about the contents of the system's file store is updated.
4. On being retrieved from the file store by the data management system, the data may be processed by a library or utility—for example, to sort it into a dictionary order, rather than the natural order imposed by the codes stored by the hardware.
5. The data management system manipulates the data, selecting those parts that the user wishes to see, and perhaps aggregating a number of distinct values into summary totals or averages.
6. An application program processes the data further, making decisions about the manner in which it should be presented to the user. For example, numbers may be presented as columns of figures, or they could be used to construct a graph.
7. Graphics services are called upon to lay the data out in the required format on a visual display.
8. A user interface manages the display that the user sees, deciding whether to present the whole of the picture produced by the graphics services, or to mask some or all of it because the user has arranged other data windows 'on top' of the area occupied by the picture.
9. Finally, the information is presented to the user.

In all of these cases, the transformation is unavoidable. It must be carried out in order to get the job done.

Clearly, the job will take longer if, in addition to these transformations, further unnecessary manipulations are performed. By the same token, the system (both hardware and software) will become larger if it has to support additional transformations beyond those which are really needed. Yet this is what a lack of standardization implies, because of the need to translate from 'foreign' data formats to those understood by a particular system, or a particular component of a system.

The traditional way to avoid this problem is to purchase as many components of the system as possible from a single supplier, assuming that the proprietary parts will work together without needing the additional data translations, or that the supplier will provide any that may be needed. While this approach may bring a feeling of comfort, it also brings constraint. New technology can be added to the system only as fast as the supplier chooses to make it available in a compatible form. The sheer pace of change in the computer industry makes it difficult for even the largest supplier to keep up without outside help.

The movement towards open systems—where ideally each of the system components may be purchased from a different supplier—allows users to keep up with technological change directly, by buying the best technology available regardless of source. However, unless there are common standards for the exchange, storage, and manipulation of data, much work is likely to be required before components can work together efficiently.

In the important area of database manipulation, standardization has been achieved through the adoption of Structured Query Language (SQL).

Structured query language

Structured Query Language (SQL), like the UNIX operating system and the Open Systems Interconnect (OSI) model for communications, is a product of the 1970s, and has taken most of the 1980s to reach maturity.

Although the relational database technology associated with SQL was originally developed by Dr E. F. Codd at IBM, it became closely linked with the UNIX operating system at the beginning of the 1980s. Essentially all of the important database management products for UNIX, and other POSIX-compliant operating systems, are relational, and all now implement SQL as a data access language. Examples of such products are Informix-SQL, Ingres, Oracle and Sybase.

Basically SQL is a specialized programming language in which data is represented as a relational model. In the relational model, data is held in two dimensional tables with a simple row and column structure.

Figure A1.6 shows the architecture of a typical implementation. Data manipulation is performed by a 'database engine' which makes underlying hardware-dependent files look like tables dictated by the relational model. The database engine also manages other files and data structures used to

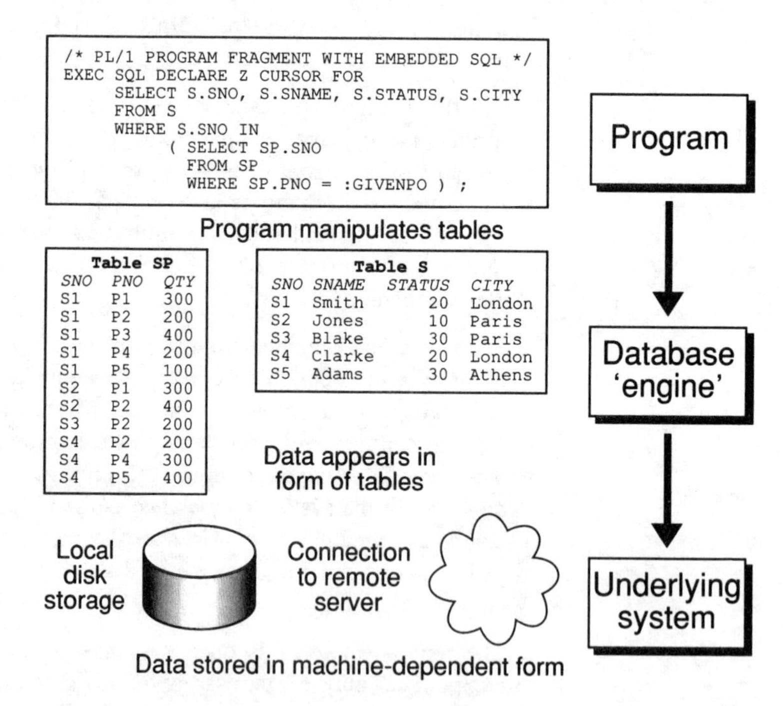

Figure A1.6 Architecture of typical SQL implementation
Source: C. J. Date, A Guide to the SQL Standard (Addison-Wesley, 1987).

speed access to the data, and to maintain its integrity. Over almost two decades, much of the development of systems which follow the relational model has been concerned with improving the speed and reliability of practical implementations.

The database engine insulates the user and programmer from the way in which the hardware represents the data, i.e. how it stores the contents of what appears to be a table. Indeed, the user or programmer is not allowed to know about the representation, and so is discouraged from bypassing the engine to get at the underlying files directly. Although directly using the underlying files might improve performance, it ties any program that uses them to a specific machine architecture. If performance is an issue it is simpler and invariably cheaper to use a more powerful processor than to change the software.

SQL's independence of data storage format brings a benefit which has grown in importance over recent years. The data may be either local, on the system running the application program which needs it, or remote, on another system somewhere else on the network. Today it is increasingly common for database management to be delegated to a central system, a server which is accessed over a network by other systems and which provides an interface to users. SQL is the language of choice in such environments.

Client-server systems of this type have brought SQL and full implementations of the relational model to personal computers. While OS/2 is capable of managing enough memory to contain the complex programs and large data buffers needed by a useful and responsive SQL-based relational database manager, the older and much more widely used MS-DOS is not. MS-DOS can, however, support application programs which access SQL engines on a remote server computer. The server may be a special-purpose proprietary system, such as those available from 3COM or Ungermann-Bass, or alternatively, it may be a general-purpose micro, mini, or mainframe computer running the server under UNIX, or a proprietary operating system.

So far, the description has drawn a tacit distinction between relational databases as a concept, and SQL as a practical means of accessing data represented in a relational manner. In defining relational concepts, E. F. Codd drew up a list of 12 characteristics which a database manager must show before it can truly be considered to be relational. These have recently been supplemented by a further three, which address issues raised when databases are distributed across several computer systems. In reality, no commercial product measures up to the original 12 criteria, although several are now very close. This means that some products are 'more relational' than others, and that none is yet as relational as pure theory would like them to be.

The mathematical relational algebra which underlies the relational model is not itself a formal standard. Standardization efforts centre on SQL. As a data access language, SQL has been criticized for ignoring or obscuring several of Codd's 12 characteristics. Despite these theoretical shortcomings, it has been widely implemented, and has consequently been standardized by ANSI (as X3.135-1986) and by ISO (as IS 9075:1987). These standards are largely identical, since ISO used the completed ANSI standard as its base document, and made few changes. ANSI based the original standard largely on a considerably cut-down version of IBM's DB2 implementation of the SQL language.

SQL was originally developed as an interactive query language to allow users to look at the contents of a database from a terminal. The interactive language has been criticized because it falls some way short of being a complete language for data manipulation. For example, it cannot manipulate strings or format reports, and has no explicit means of making decisions based on data values (that is, there is no 'IF' statement). Consequently, SQL has been standardized as an embedded language and the interactive aspects of SQL are not standardized.

Embedding is a process whereby fragments of SQL appear in a program written in some other computer language. The SQL statements manipulate the database. The surrounding program manipulates data to be stored and data which has been retrieved. The ANSI standard for SQL deals specifically with the cases of COBOL, FORTRAN, Pascal and PL/I, all languages for which ANSI standards exist. Embedding techniques (known as *bindings*) for newer languages like Ada and C have not yet been standardized, although in practice there is little variation between different suppliers' implementations. The X/Open Portability Guide has defined a *de facto* standard for C bindings since 1987.

It is worth noting that all of the languages for which formal bindings exist are third-generation languages. Most fourth-generation languages embody powerful facilities for database access. Some use SQL as their primary (or indeed only) tool for this purpose. Others, as a result of customer demand, offer SQL as an alternative to their intrinsic tools. Where SQL is provided facilities may be restricted to, for example, read-only access to data. Since fourth-generation languages are not yet standardized, it follows that the use of SQL within such languages cannot be standardized.

The formal standards for SQL are minimal. They define a limited set of requirements to which existing SQL-based products can conform easily. Importantly, the minimal set does define transaction processing constructs. These allow a sequence of operations to be bracketed together in a way that ensures either that all of the operations are carried out in full, or that, in the event of an error at any point in the transaction, none of the operations takes place. Without this facility, it is impossible to maintain the integrity of the database in the face of hardware, software, or operator errors. Delay in incorporation of the standard's transaction processing constructs has been the main reason for the lack of complete conformance to the ANSI standard among otherwise satisfactory products.

Strictly speaking, the description of the public standards as minimal refers to their *level 1* requirements. A second level is also defined, which tightens up and extends the requirements of level 1. While still quite minimal, level 2 did not match the features of any one SQL-based product at the time the original

ANSI standard was released, and still does not. Generally, the limits are of a practical nature and not necessarily related to Codd's criteria. For example, it is a practical necessity to be able to delete unwanted tables, yet the DROP statement used by most implementations is not covered by the standard.

A program which confines itself to level-1 features of the standard will be maximally portable. But level 1 is so limited that there is often a need for the programmer to use features beyond it. Even where the constraints can be observed, in writing a program which accesses an existing database, features outside level 1 (and, quite possibly, outside level 2) will be needed to set up and administer the database. This is particularly true of the transaction processing constructs. The standard does not address the topic of rebuilding a database following a serious failure.

ANSI and ISO are both working on more complete standards for SQL, but no results are expected before 1991. In the mean time, those requiring a more comprehensive, if less definitive, working definition should refer to the X/OPEN Portability Guide (XPG). Figure A1.7 illustrates the relative scope of ANSI X3.135-1986 and Issue 3 of the XPG.

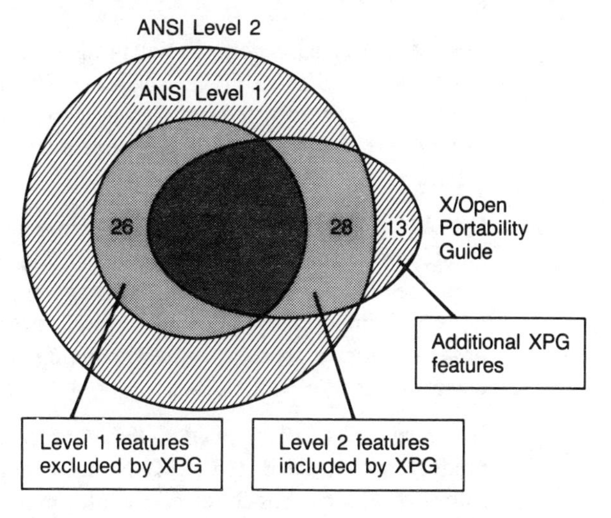

Figure A1.7 Scope of SQL standards

X/Open's definition relaxes a few of the requirements of level 1 of the ANSI standard. For example, ANSI defines a format for comments in SQL programs while X/Open does not. In most respects, however, X/Open is stricter than level 1 of the ANSI standard, incorporating many aspects of level 2, and a number of extensions beyond the standard which are of a practical nature. For example, X/Open requires a DROP statement, and defines C language bindings.

Although no supplier yet (1990) provides an implementation of SQL which fully conforms to the requirements of the XPG, several are expected to appear by 1991.

Index-sequential access method—ISAM

Up to this point, discussion has centred on relational database management systems, tools capable of manipulating and correlating data in a very complex manner. Many applications do not require this power—they could function quite adequately without the expense of the additional memory (software costs) that are the price of using a full relational database manager.

Index-sequential access methods (ISAMs) answer this need and predate database management packages. An ISAM simply allows a record to be fetched from a file by reference to a key field which is part of the record. Similarly, records can be deleted or updated by using their key, and, if desired, new records can be added. The COBOL language has incorporated facilities for indexed access to files since the mid 1960s.

The *de facto* ISAM standard for open systems is a proprietary product, C-ISAM, from Informix Software Inc. C-ISAM was one of the first commercial add-on products for the UNIX operating system. It was introduced in the early 1980s and was used to add indexed file access facilities to the operating system itself.

Since its birth as a commercial product, C-ISAM has become a *de facto* standard for MS DOS, OS/2, UNIX and Digital's VMS operating systems. It has been widely imitated by other software vendors, in part to improve performance but also so that developers could avoid royalty payments to Informix.

C-ISAM, which is a library for the C programming language, was also incorporated into another proprietary product, Micro Focus Level II COBOL. This implementation of the COBOL language opened UNIX to business applications and, importantly, allowed COBOL programs written for Micro Focus' original product base, personal computers, to be ported to the UNIX environment.

Informix Software followed C-ISAM with Informix, a relational database product which used C-ISAM as an underlying file-access mechanism. Informix, in turn, was followed by Informix-SQL, a database manager which implements Structured Query Language, but which can still use C-ISAM to access files.

Limitations in the performance and generality of the ISAM approach have resulted in its replacement with an alternative access method. Consequently, the recent Informix-Turbo variant of Informix-SQL does not use C -ISAM. These limitations are important not just for database management packages, but for any application where use of an ISAM might seem appropriate, and are discussed below.

Standardization status

In 1987, X/Open included a definition of C-ISAM in issue 2 of its Portability Guide, calling the C language library 'X/Open ISAM'. Although based on Informix' own specification, the XPG's definition omits a few features which were judged to be implementation-dependent. Among these is an audit trail mechanism which could be used to recover modifications to data file records following a system failure. X/Open ISAM interfaces are included with several recent implementations of the UNIX operating system.

Issue 3 of the XPG implies, but does not actually require, that random-access files used by COBOL programs should also be accessible to C language programs through the ISAM facility As the XPG augments the ANSI standard (X3.23-1985) with features derived from Micro Focus Level II COBOL, it is likely that XPG-conformant COBOL support systems will allow this type of access.

X/Open's commitment to the C-ISAM interface marks the library as a well-established *de facto* standard, but, despite lobbying by X/Open and others, no public standards body has yet shown interest in basing a public standard on X/Open's ISAM work. This lack of interest perhaps comes about because ISAMs represent old technology. All the public standards activity in the field of data access concerns relational database technology.

ISAM and relational database compared

A relational database manager can do anything that an ISAM can do, and more. Figure A1.8 summarizes the relative merits of the two approaches.

The initial and continuing costs of a relational database manager are higher than those for an ISAM, both in terms of cash, and of the amount of computer memory that is consumed. However, these investments bring far more flexibility than can be achieved with an ISAM, and are likely to cut programming costs and to improve the speed of data access in all but the simplest applications.

Programming costs are reduced because SQL is a high-level language for data manipulation, whereas the functions in an ISAM library provide only for low-level operations. For example:

- A relational join between two database tables can be accomplished with a short SQL command. To accomplish this, common function using calls to the ISAM library takes many lines of program code.
- SQL can calculate averages and totals. An ISAM library cannot, with the result that program code must be written to do the job.

All commercial vendors of SQL devote research and development effort to the optimization of their products. The result is that complex operations are likely to execute more quickly than the corresponding series of low-level calls to an ISAM library. This improvement in execution efficiency is likely to compensate for the overhead involved in the start-up of the SQL support system. Only a very competent ISAM programmer is likely to be able to improve the performance of a commercial database manager in the general case, although there is a place for ISAM in small, specialized applications.

System configuration independence is another important issue. In particular, commercial SQL database products and the public standards for SQL are

	ISAM	*SQL Relational Database*
Type of standard	*De facto*	Public
Initial cost	Zero to low	Medium to high
Sub-licensing cost	Zero to low	Low to high
Memory requirement	Low	High
Efficiency (simple jobs)	Medium to high	Low to medium
Efficiency (complex jobs)	Low to medium	Medium to high
Program complexity (simple jobs)	Medium	Low
Program complexity (complex jobs)	High	Medium
Computational ability	No	Yes
Transparent access to distributed database	No	Possible (implementation-dependent)
Support for interactive *ad hoc* queries	Vendor-dependent; requires fore thought	Yes
Reconfiguration of database	Requires custom-written program, modification of dependent programs	Easily achieved with SQL; dependent programs unaffected
Recovery after failure	Requires forethought and custom-written program	Requires forethought; further action automatic in some implementations
Transaction processing facilities	Non-standard option	Required by standard
Standard languages supported	C, COBOL	COBOL, FORTRAN, Pascal, PL/1
Additional languages supported	BASIC, Pascal (language vendor-dependent)	C

Figure A1.8 Comparison of ISAM and relational database manager

proceeding in the direction of distributed databases. These are databases where information is distributed across a number of systems, connected using local or wide-area networks. Just as the applications which access a relational database do not need to change if the structure of the underlying database is altered, so they do not need to change when the database is distributed. This is not true of ISAM-based applications, which must incorporate knowledge of the location and format of the data. Consequently, while remote file access mechanisms may allow an application to use ISAM files held on a remote system, any application which does this must incorporate information about the configuration of the network. It will be difficult to maintain if the network configuration varies, as it is likely to do if the application is installed at a number of sites.

Efficiency is also an issue for distributed databases. Figure A1.9 shows that a distributed database, where a server program on a remote machine acts on

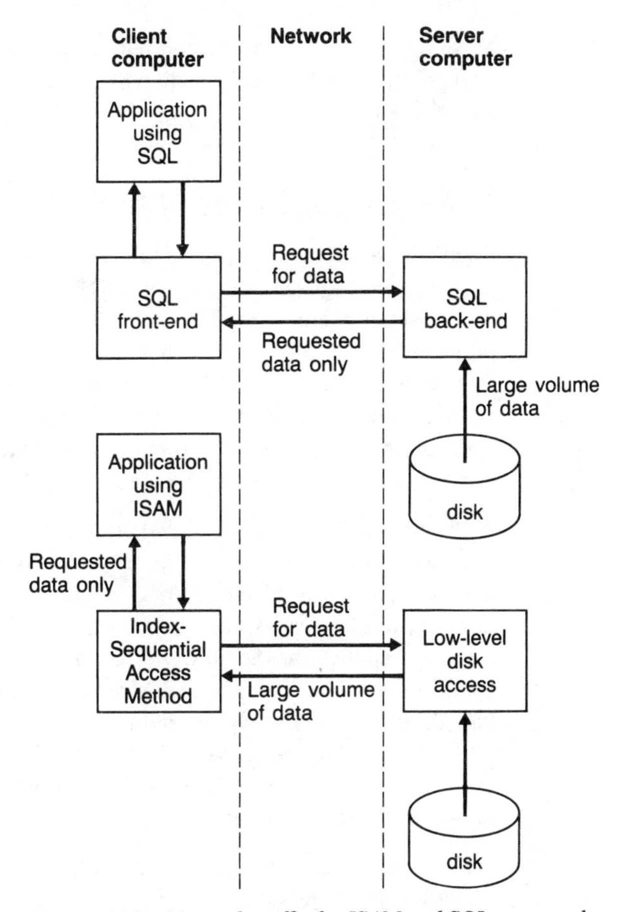

Figure A1.9 Network traffic for ISAM and SQL compared

requests from an application, can result in lower network traffic than a more traditional arrangement, where large amounts of data may cross the network. This is achievable because the network only passes the results, not the total data from which the results are derived. Any reduction in network traffic is desirable. The server on local-area networks can often be a bottleneck, and the transport of high volumes of data across wide-area networks is likely to be time-consuming.

Problems for existing applications

The large base of existing applications using ISAM, written in C or COBOL, can be difficult to convert on to distributed systems. There is no automatic means of translating low-level ISAM file manipulations into high-level embedded SQL. The affected programs must be extensively rewritten if they are to be able to take advantage of the new environment. Cost considerations make it unlikely that this will be done. As a result, ISAM is likely to be required for many years.

ISAM: a summary

Technical considerations mean that the use of an ISAM is seldom preferable to use of a relational database in new applications. Cost considerations may make an ISAM more attractive, but savings in licence payments must be set against increased development and maintenance costs, and limitations in future flexibility.

Fourth-generation languages

Fourth-generation languages (4GLs) evolved from 'high-level' third-generation languages, such as COBOL, FORTRAN, and PL/1. They appeared commercially in the early 1970s. Just as third-generation languages promised higher productivity than the assembly language and machine code that they replaced, so 4GLs were promoted as a means of reducing the amount of time needed to write computer programs. This was expected to cut application backlogs within organizations, and allow software developers to bring products to market more quickly.

As Figure A1.10 illustrates, the evolution of 4GLs is one aspect of the developments arising from the third-generation languages. In fact, the third generation really consists of two distinct components. The earlier component, the unstructured languages, has produced FORTRAN (predominant in scientific computer applications), and COBOL (the most popular language for business applications). The second, block-structured languages, includes Algol, C and Pascal, and has given rise to the newer, object-orientated languages such as C++, Modula 2 and Smalltalk. Some of the ideas developed in block-structured languages have also found their way back into the current versions of unstructured languages.

4GLs can be considered to have evolved from COBOL, in that almost all of the products on the market are designed for the rapid production of business applications. There are no significant 4GLs which replace FORTRAN as a scientific language, or C as a systems programming language. 4GLs improve on COBOL in a number of ways:

- Where an actual written computer language is involved, the language is more powerful and less verbose than COBOL.
- Some 4GLs remove the need for a written language completely. An application is created by filling in forms on a screen, directly creating a machine readable description of its logic and of the files that it manipulates.
- 4GLs are usually associated with databases, making the language closer to the information that it must manipulate than, for example, is COBOL. Many 4GLs actually build up the description of an application by creating entries in a database. Early 4GLs were specific to particular database managers (often supplied only by the supplier of the 4GL) but today's most successful products allow the user a choice of database software.
- 4GLs embody constructs which facilitate the rapid production of screen forms and reports. Recent product releases also allow overlapping screen windows to be manipulated by applications. Standard COBOL lacks these interactive features. They may be obtained for COBOL through teleprocessing monitors, which tie the resulting program to a proprietary architec-

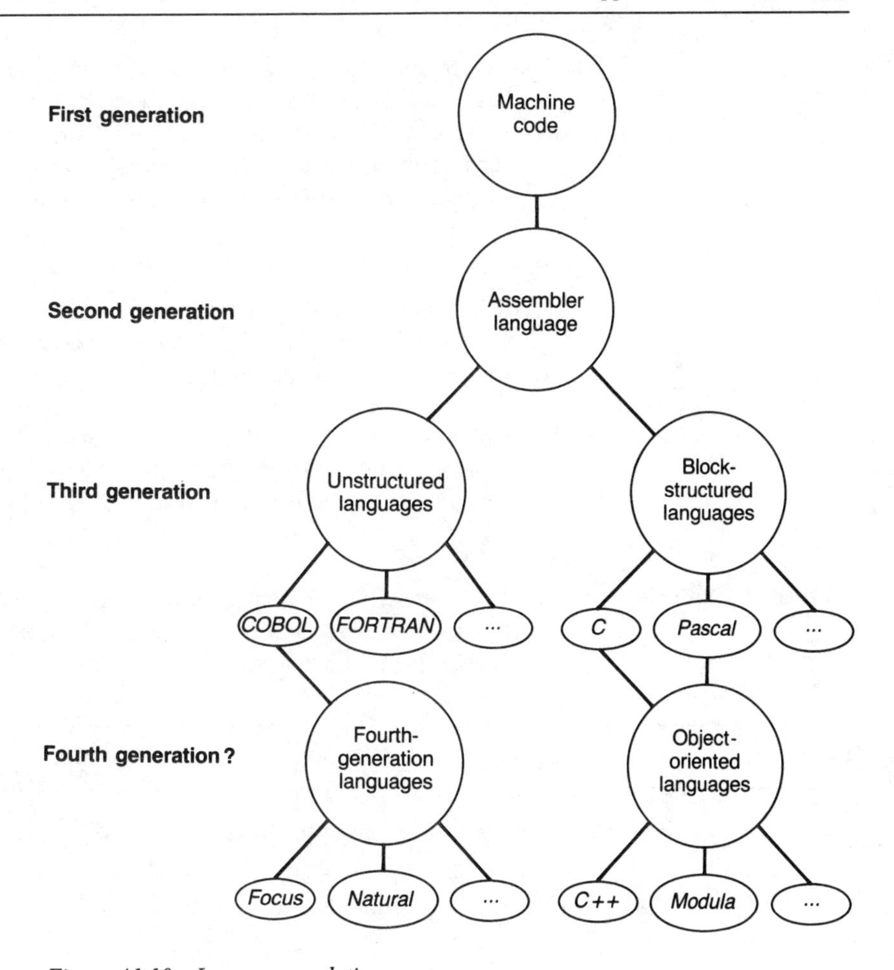

Figure A1.10 Language evolution

ture, or through screen-handling extensions such as those specified in the X/Open Portability Guide, but this can result in very long programs.

- 4GLs can give a large measure of machine independence, spanning a wide range of machines from personal computers, through minis, to mainframes. This brings a high level of program portability across these platforms, and provides access to a variety of database management systems.

The last point is of particular interest to advocates of open systems. 4GLs can bring independence from a particular hardware architecture, although possibly at the cost of commitment to a single software supplier.

Until recently, very few of the most popular 4GLs from the mainframe and minicomputer worlds were available for the UNIX environment, although a few had been implemented (often in cut-down form) on the IBM PC. The UNIX world had a number of fourth-generation tools, for example Accell (Unify Corp.), Informix-4GL (Informix Software Inc.), and Progress (Progress Software Corp.). Towards the end of the 1980s, however, suppliers of 4GLs for proprietary architectures began to enter the UNIX marketplace, both because they judged that sales volume would be sufficient to support the heavy conversion costs and because their customers were demanding support in the UNIX environment. As a result, tools from 'traditional' 4GL suppliers, such as Information Builders (Focus) and Software AG (Natural) are now available for the UNIX operating system. This makes open hardware environments an option for organizations committed to these products in their mainframe and minicomputer operations.

It is interesting that many suppliers of 4GLs are seen primarily as database suppliers: Informix, Oracle, Relational Technology, Software AG and Unify, for example. This is not surprising, as 4GLs are usually heavily dependent on database facilities. Other 4GL suppliers, although they can provide database facilities, make a point of their independence. Examples here are Information Builders and the SAS Institute. Even Unisys is promoting Mapper, a product which was originally both hardware and database-specific, as a independent 4GL.

Standards status

There are as yet no standards for 4GLs, nor is there any prospect of standards, either *de facto* or public, for the foreseeable future. Figure A1.11 summarizes the position, and compares it with that for earlier generations.

Generation		*Language type*	*Population*	*Standards*	*Examples*
1st	1940s	Machine codes	Mostly hardware	Some; *de facto*	specific
2nd	1950s	Assembly languages	Mostly hardware specific	None	Intel 8086
3rd	1960s	Third-generation languages	Mostly; application area-specific	Many; public	Ada; C; COBOL; FORTRAN; Pascal; PL/I
4th	1970s	Fourth-generation languages	Many; mostly for business applications	None	Easytreave Focus Mapper Natural

Figure A1.11 Language generations and standards status

There are, perhaps, two major reasons that the situation differs so markedly between third- and fourth-generation languages:

- Third-generation languages are older—in some cases, much older—than fourth-generation tools. Consequently, there has been more time for the standards process to work.
- In general, each major third-generation language has established supremacy in a particular niche: COBOL for business use, FORTRAN for scientific, and so on. This natural selection marks out clearly the languages which should form the basis of standards. In contrast, all 4GLs are essen-

tially competing for a single niche—that of the successor to COBOL—and no clear winner has emerged.

There has been one move to standardize a language which could be described as fourth-generation. Ashton-Tate's dBASE product incorporates constructs for database access, report generation, and screen manipulation, and has proved to be a powerful productivity tool in the hands of experienced users.

In late 1988, a working group (1192) of the IEEE announced an intention to draw up a public-domain specification for the language, based on publicly funded work at the Jet Propulsion Laboratories, done before dBASE became a commercial product. However, the project was frozen when Ashton-Tate stated that it regarded the language as proprietary, and would not cooperate with any move towards making it a public standard.

Since similar reactions could be expected if an attempt were made to standardize any other 4GL, it seems unlikely that standardization will take place in the near future.

Computer-aided software engineering (CASE)

Computer-aided software engineering (CASE) tools are often promoted as offering productivity gains similar to those which can be achieved with 4GLs. The tools automate the mechanical aspects of the construction and maintenance of programs, using computer power to help the programmer to apply one of a number of techniques developed to speed production. Importantly, CASE tools can also keep track of the modules which make up a program, and the programs which make up an application, so helping with the serious problem of maintaining large suites of application software.

Originally developed for third-generation languages, CASE techniques can also be used with some 4GLs. Indeed, over the last couple of years, several established 4GL products have been augmented with CASE subsystems. Examples are Oracle's CASE*Designer and Software AG's Integrated System Architecture for the Natural language.

There are many public standards for software engineering. In particular, the IEEE has developed over a dozen linked standards since 1983, with more in the pipeline. All approved IEEE standards are also ratified by ANSI. This effort is being coordinated with ISO, although no international standards have yet been released.

It is important to realize that the standards that do exist relate simply to software engineering—not to computer-aided software engineering. Commercial CASE products seldom implement public standards, but instead automate less formal procedures described in the literature and widely used in commercial programming. Examples are the Jackson Method and the Structured Systems Analysis and Design Method (SSADM). To these are added database tools which keep track of the subsystems which make up an application.

While it does not embody anything that could reasonably be called CASE the UNIX operating system has supported tracking tools of this type since the mid 1970s. Examples are: *make*, which rebuilds complete applications from their components; and *rcs* (Revision Control System) and *sccs* (Source Code Control System), which keep track of changes to program source codes. All three utility packages are widely distributed both across implementations of the UNIX operating system, and in several other environments.

There is a high degree of commonality between implementations, marred by small differences. As a result, programmers wishing to use these tools in a portable manner must be aware of the features which may safely be used. Standardization is required, and is in the province of the IEEE's POSIX (1003.2) working group, which is already committed to other activities until the early 1990s.

Summary: data handling

- Structured Query Language, SQL, is widely accepted as an international standard for database access.
- There are usable standards for the embedding of SQL in the popular third-generation languages, such as COBOL and FORTRAN.
- No database products conform completely to existing standards. The standards themselves are incomplete in some respects. Consequently, it is difficult to write useful programs that are portable across all implementations of SQL.
- Current standards do not address the issue of distributed databases, where data held by several computers appears to the user as a single resource. Several commercial products provide this facility, and standards are under development.
- The low-level standard for an ISAM promoted by X/Open is widely accepted, but is not currently scheduled to become a public standard.
- Fourth-generation languages are increasingly used as a more productive alternative to third-generation languages, but there is no clear market leader which might become a *de facto* standard.
- CASE tools exist that provide for increased productivity in software development, but there are no standards, *de facto* or public, in this area.

Appendix 2 The user interface

In Chapters 5 and 8, we saw that the user interface is a very important component of open systems. People who use computer systems and software would like to have a large measure of commonality in the interfaces of the applications software products that they use. This would reduce the need for retraining when a user familiar with one package is introduced to another; knowledge of the first package could be re-used in becoming familiar with the second because both have a similar 'look and feel'. At the same time, a wide selection of applications that provide the same 'look and feel' is necessary.

Commercial software developers have until recently resisted the imposition of a particular user interface style, arguing that a constraint of this type could limit programmers' ability to differentiate their product from its competitors.

There is some merit to these arguments in the case of traditional character-based terminals. Graphics displays, however, allow many elements of differentiation above and beyond a standardized user interface. Add to this the 'commodity' status of graphics displays and it becomes very difficult to sustain any negative arguments about a standardized graphics interface.

In this appendix, we review current developments in standardization of the graphical user interface.

User interfaces: a technical overview

From the point of view of standardization, it is important to realize that several software subsystems must work together in order to provide a practical user interface. Figure A2.1 shows a system component model which gives one perspective of the user interface.

The user interface component appears to be at the heart of the issue. In the past, computer systems have not traditionally provided an explicit and identifiable user interface component for use by applications. Instead the job of implementing an interface was left to each program, possibly with help from subordinate system components.

Graphics services play an important role in modern user interfaces. They provide a mechanism which allows programs to manipulate and draw arbitrary shapes and patterns on a screen or on paper. Graphics services also provide a means for the user to convey information about position and movement to an application, typically by means of a mouse, track-ball or graphics tablet. Characters, of course, can still be entered with a conventional keyboard. This graphical interface contrasts with the older, character-based

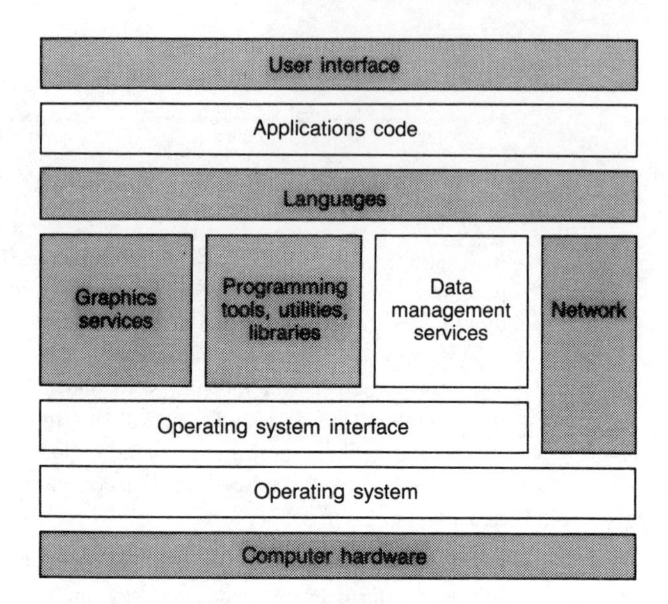

Figure A2.1 Subsystems involved in user interface

approach, in which characters formed the only means of communication between an application and its user.

Libraries are often used to provide graphics services to those applications which need them. Environments such as Apple Computer's Macintosh operating system and Microsoft's OS/2 provide such services from within the operating system, because almost all applications are expected to require them. Libraries also provide services to character-based applications—for example, to make the application independent of the type of computer terminal chosen by the user.

Network services are also implicated in user interface issues. This comes about because the user's terminal is often remote from the computer which runs the application program. In the past, low-speed communications schemes moved character-based data between the computer and the terminal. Now, with more modern graphics-based schemes, high volumes of data are moved quickly and reliably, often using local-area network technology.

Computer hardware must provide an interface to the user's terminal. This low-level interface should not be used in a particular program, or it will be difficult to move the program to other types of hardware. Almost all graphics software written for the IBM PC suffers from this problem, making the applications difficult to port to other environments, even those capable of running on identical hardware, such as OS/2.

A different perspective of the user interface components is given in Figure A2.2. As with the OSI model for communications, a multi-layer model has been developed to describe the components of user interface systems, although, unlike the OSI seven-layer model, this does not yet have backing as a formal standard.

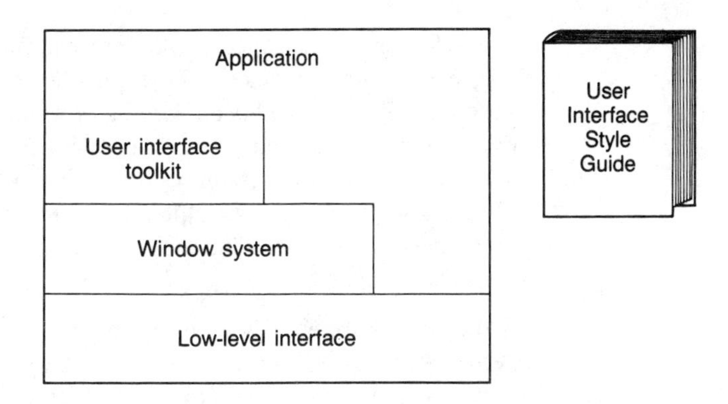

Figure A2.2 Multi-layer model for user interfaces

An application program can access the user interface by using powerful tools from a *user interface toolkit.* These tools, for example, create and manage pop-up menus, windows with scrolling and resizing facilities, consequently creating a particular 'look and feel'. All menus created with the toolkit will have a similar appearance and all windows created have similar facilities for scrolling and resizing. If several applications use the same toolkit, then those applications will share the same 'look and feel', making it easy for users familiar with one of the applications to use the others.

The user interface model allows programmers to circumvent the 'look and feel' imposed by the toolkit if they wish by making direct use of the *window system* which underlies it. The window system implements the basic features necessary for a graphical user interface: writing text and graphics into rectangular windows on the screen; keeping track of the mouse position, and so on. However, it does not restrict the programmer as to the manner in which these elements are put together. This means that a choice of user interface toolkits can be offered on top of a single window system, and that an application can implement its own unique interface if it wishes.

The *low-level interface* forms the base of the model. Again, it is possible to offer a choice of windowing systems on top of a given low-level interface. An application program can access the low-level interface directly if it wishes. The vast majority of applications should not need to, provided that the facilities provided by the higher-level libraries are adequate. Instead, direct access should generally be confined to programs which administer the user interface. Incorrect use of the low-level interface by one program can easily cause reliability or security problems for other programs with which it is sharing a display device.

The final part of the model is not a software component, but a document—the *style guide*. The purpose of the style guide is to describe the components and procedures which together make up the 'look and feel' of a particular user interface. While the choices a programmer has in using a user interface library are constrained, there is still some flexibility. For example, two applications may share common features, such as the ability to cut, copy, and paste information displayed on the screen. But if one chooses to activate these features through function keys, and the other through menu options, the similarity is masked, and users familiar with the first application may find it difficult to use the second. Even if both applications use menu options for the facilities, user confusion could result if one application calls the 'Copy' operation by that name, while the other names it 'Duplicate'. The style guide tells programmers how to present such facilities in a manner which is common across applications.

Style guides are often bulky—for example, AT&T's guide for its Open Look interface runs to 250 pages—and are not necessarily read and adhered to. As a result, applications can be produced which do not completely conform to a particular 'look and feel', even though they have used the correct user interface library.

Given the choice, many programmers seem to prefer to use lower-level facilities such as those of a window system, rather than those of a higher-level toolkit, on the basis that lower-level functions are more flexible. This may be correct, but one goal of a toolkit is to constrain the programmer so that a more consistent user interface can be presented. A second argument used is that lower-level functions are likely to be more efficient. This has been proved incorrect.

Figure A2.3 relates system components to the user interface model, illustrating that the low-level interface does not correspond to hardware. Recently developed user interfaces hide details of the underlying hardware from the application, both to promote portability and to mask the differences between integrated displays and keyboards and remotely connected devices. The interface allows application programs to enquire about the characteristics of the

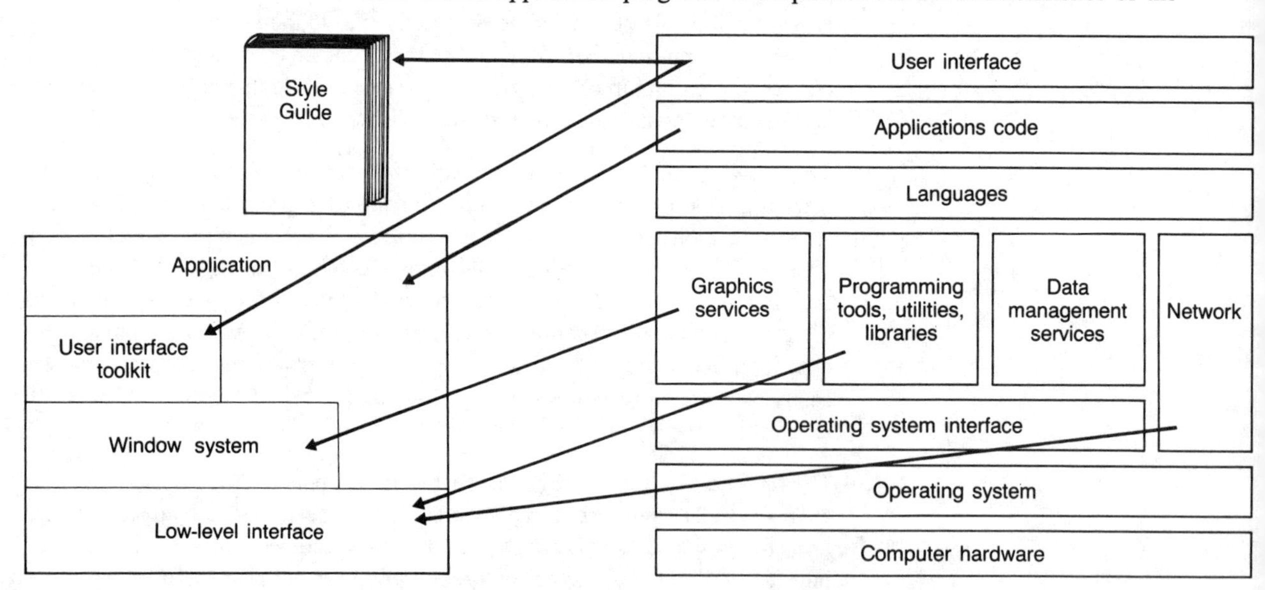

Figure A2.3 Relationship between user interface model and system component model

connected devices—the resolution of the screen, whether colour is supported, the characters and functions supported by a keyboard, and so on—and adjust itself accordingly. In this way, the same program can have a number of different devices connected at the same time (or even from time-to-time) and adjust itself accordingly.

Character-based user interfaces

Standards for character terminals—and why they are ignored

Until the early 1980s, computer systems for commercial applications had character-based interfaces. Only the more expensive scientific and technical equipment provided the means to display graphical output and accept graphical input. Character-based screens were (and are) typically capable of displaying 24 or 25 rows, each containing 80 characters. Each of these 1920 (or 2000) character positions is addressable, meaning that the computer can update the character displayed at any position without affecting the data elsewhere on the screen.

In 1979 an ANSI Standard (X3.64) covering the addressing of the screen and additional functions such as bold or reverse-video display of data, was agreed. The 1988 international ISO standard (6429:1988) adds a number of extensions, but does not yet command widespread conformance. It can also be argued that it has arrived too late, since the focus of development interest is now graphical displays, rather than character displays.

Although ANSI X3.64 broke new ground when it was written, it must today be regarded as a minimal standard. Those terminals which adhere to it—most notably Digital Equipment's VT100 (and compatibles) always add proprietary extensions. These non-standard extensions are undeniably useful, giving advantages such as the ability to insert and delete lines and characters without the need to rewrite the whole screen; additional display characteristics, e.g. as colour and line-drawing character sets; and the provision of function keys. Digital's own extensions have become widely accepted as *de facto* standards, but implementations vary and there is no formal test for conformance. System software supplied with the IBM PC makes its screen emulate that of an enhanced ANSI standard-compatible terminal.

While conformance to the minimal ANSI standard is widespread among the suppliers of character displays, it is far from universal. There are two major reasons for this. The first is that equipment suppliers have continued to use non-conformance as a means of locking customers into the use of a terminals which use a particular proprietary control scheme (or to computer equipment or software which requires the use of a particular scheme). Terminal specialists such as Lear-Siegler, Qume and Wyse have been notably successful over the last ten years, selling many millions of non standard-conforming devices at attractive prices. Among computer manufacturers, Hewlett-Packard supplies an extensive and popular range of terminals which use a proprietary control scheme.

While the first reason for non-conformance is proprietary lock-in, the second reason is technical. The ANSI standard covers only the character-at-a-time (asynchronous) terminals that have traditionally been used with mini-computer systems. Such terminals transmit characters to the computer as soon as the user presses a key on the keyboard, and update the screen as soon as a character is received from the computer. This scheme allows comparatively simple (low-cost) electronics to be used in the terminal, but places a requirement for the rapid processing of individual characters on the computer to which it is connected.

Mainframe computers have tended to use a different type of terminal, (known as block-mode or synchronous), which is more complex, and which is designed to cut down the load on the central computer. With these terminals, the display on the terminal's screen is updated when the user hits a key, but no data is sent to the computer. Only when a 'transmit' or 'send' key is pressed is the information on the screen (or selected parts of that information) transmitted to the central computer as a single block. This arrangement works well for typical business applications (in particular, transaction processing) but is less suitable for today's highly interactive applications such as word-processing with automatic line-wrap and hyphenation.

The major *de facto* standard for block-mode terminals is derived from IBM's 3270 series (for example, the 3274 terminal cluster controller; the 3279 colour terminal). Many 'plug-compatible' suppliers market synchronous terminal equipment which may be substituted for IBM's. Other suppliers of mainframe computers, such as Control Data and Unisys, use their own proprietary schemes for the control of synchronous terminals.

Coping with the lack of standards

Multi-user open operating systems, most notably UNIX, were initially developed on mini-computers, and so were designed with character-at-a-time terminals in mind. Mainframe implementations of UNIX initially supported only synchronous terminals, but customer pressure has resulted in the provision of support for asynchronous terminals as well.

It is reasonable to expect that an open operating system which runs on a variety of types of computer hardware should also support a variety of types of terminal, and, importantly, should provide a means whereby application programs can be made independent of the type of terminal chosen by a particular installation or user. This situation differs from that in a closed environment, where applications need to cope only with the type of terminal supported by the environment. Terminal support in a closed environment is easily achieved by coding details of the terminal into the application program.

Figure A2.4 shows an example of an application which needs simply to clear the terminal screen, and to respond to a function key. The program itself contains the control codes required. Many character-based applications for the IBM PC are written in this way, using 'wired-in' ANSI standard sequences to control the screen.

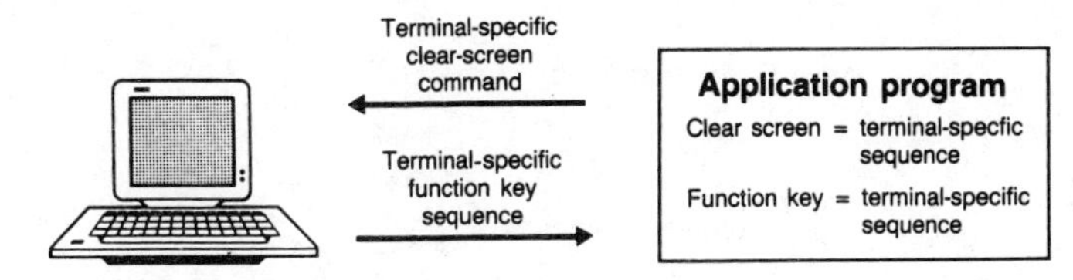

Figure A2.4 A terminal-specific program

Parts of the program must then be rewritten if alternative terminals are to be handled. This does not matter on a personal computer, where the terminal is, in effect, part of the computer and has fixed characteristics. But when an attempt is made to move the application on to other types of machine, such as a multi-user computer, the problems become apparent immediately. In order to gain porting flexibility, software authors learned to write applications as shown in Figure A2.5, where applications can retrieve the control information needed by particular types of terminal from a specialized database.

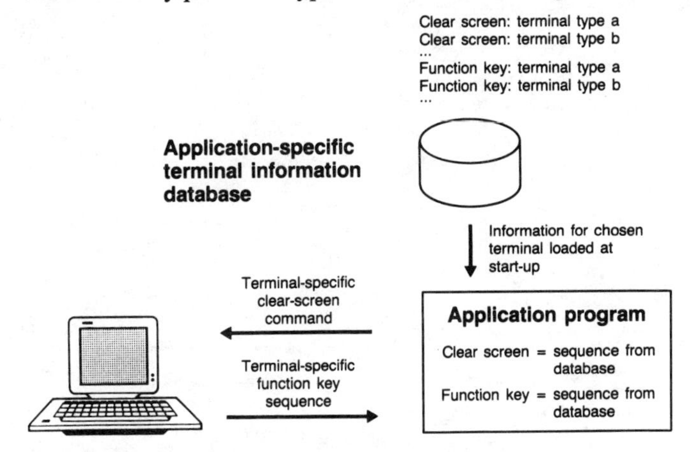

Figure A2.5 A terminal-independent program

This arrangement allows new types of terminal to be supported simply by adding information to the database. It is no longer necessary to amend the application software. However, the approach has a drawback. If a computer system is host to several applications from a selection of software authors, terminal information databases may proliferate, as Figure A2.6 illustrates.

This presents a maintenance problem for the system administrators, who must update each database whenever a new type of terminal is introduced. Figure A2.7 shows a better solution, which involves the provision by the system supplier of a library for terminal control.

Not only does this ease the task of administration but a comprehensive library should also remove the need to recreate detailed low-level terminal control logic for each application.

De facto terminal handling standards in UNIX

The UNIX operating system has long supported a mechanism of this type, in the form of a library called *termcap*. This is used in conjunction with a supporting database containing descriptions of all the terminal types known to

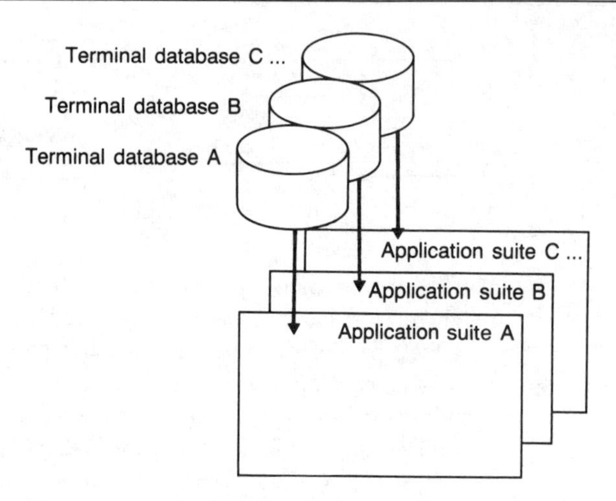

Figure A2.6 Terminal database explosion

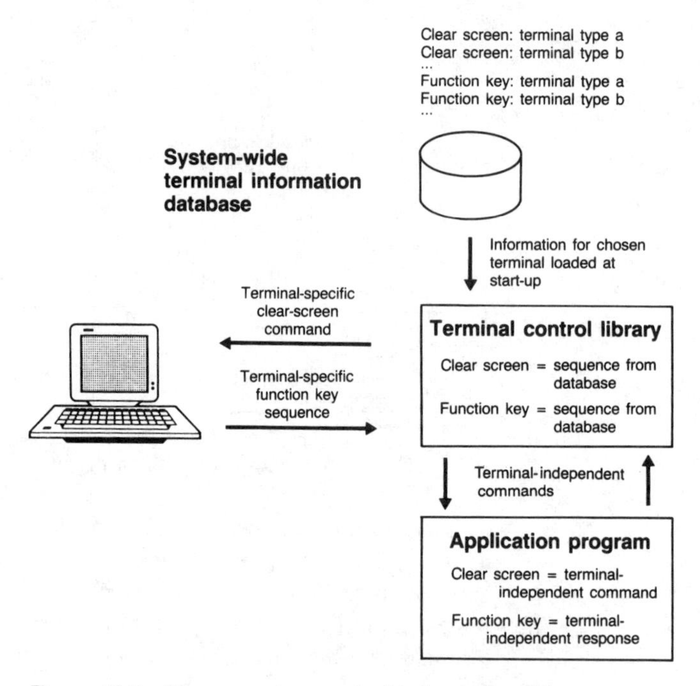

Figure A2.7 Program using terminal-independence library

the system. Termcap is a low-level library. It provides a means of access to the individual capabilities of a terminal—clearing the screen, inserting characters, and so on. It forms the basis for *curses*, a higher-level library which allows programs to treat the screen as a series of windows containing characters. Curses looks after the details of translating the program's manipulation of the data in each window into a sequence of low-level operations which update the display on the screen.

The original termcap scheme provided a worthwhile solution to the problem of terminal handling, but had a number of technical shortcomings.

Consequently, AT&T decided to replace it with a new low-level terminal information library, *terminfo*, in UNIX System V. A version of curses based on terminfo appeared at the same time. In order that System V could support existing applications and source codes, AT&T continued to ship termcap *and* the corresponding version of curses in addition to their replacements. Also, since one of the differences between terminfo and termcap concerns the format of the terminal information, two databases of terminal information were also shipped. Figure A2.8 shows the current status.

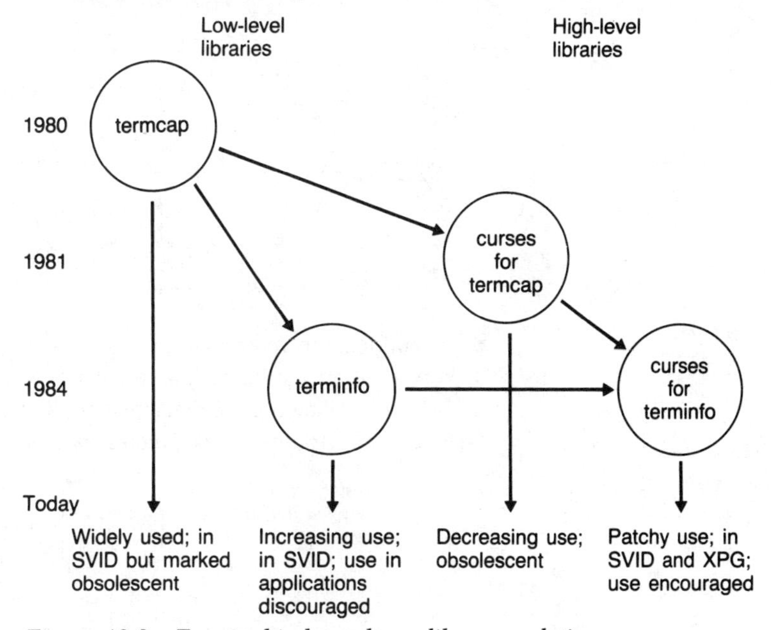

Figure A2.8 Terminal independence library evolution

None of the mechanisms described has reached public standard status, nor has any standards body adopted any of them as a draft standard. Only curses have firm status as a *de facto* standard, being backed both by AT&T's System V Interface Definition (SVID), and X/Open's Portability Guide (XPG). Terminfo, while defined in the SVID, is not included in the XPG. This means that XPG-conformant environments can use any low-level terminal control mechanism that they choose. This mechanism may or may not be visible to applications programs.

It seems, then, that curses should be the tool of choice for portable applications which must interact with their users through character terminals—a description which covers a large proportion of the applications in use today. Sadly, this is not the case. Many applications use termcap, terminfo or curses and modify the information in the system-wide database by one of several methods.

In one such method, as Figure A2.9 shows, additional application-specific information is added to the system-wide terminal information database. The termcap database is particularly susceptible to this type of amendment, because it does not support useful terminal features such as box drawing character sets, colour, and large numbers of function keys. Applications wishing to access these features could do so only by supplementing the information held by the system. Inevitably, some combinations of applications require conflicting changes to the database, an administrative problem which can be solved only by creating private copies of the database for problem packages.

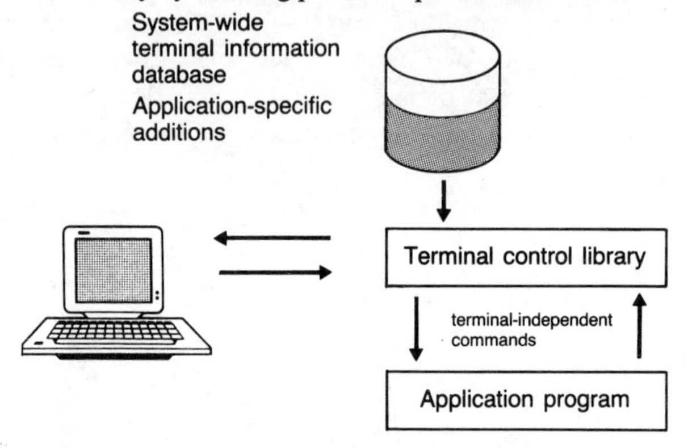

Figure A2.9 Application-specific additions to system database

Some applications require no changes to the system database, but supplement it with additional information stored elsewhere, as in Figure A2.10. While slightly better than the previous method, this mechanism still requires that the system administrator maintains more than one database.

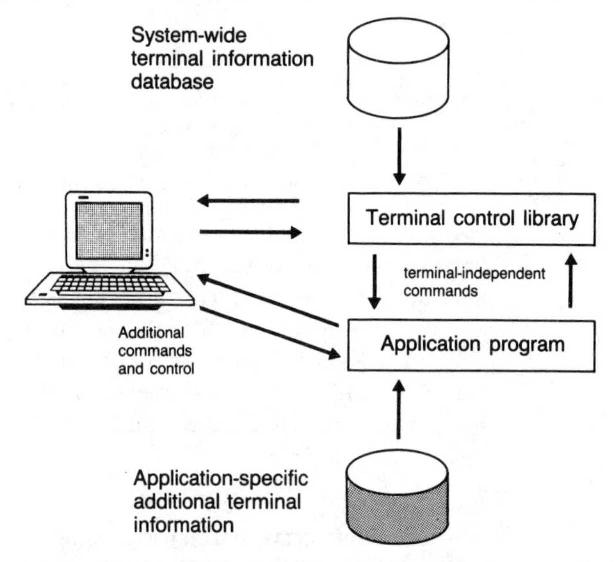

Figure A2.10 Additional information in application-specific database

Why has this unsatisfactory situation come about? The primary reason is that the capabilities accessible through the UNIX libraries and their supporting databases have, until very recently, lagged behind the capabilities of commercially available terminals. For example, terminfo did not incorporate support for box drawing until 1987, or for colour until 1988. (Neither of these capabilities is yet defined in the SVID, or in the X/Open Portability Guide.)

Consequently, developers needing those facilities had no option but to extend the library mechanisms in some way. Once such extensions have been incorporated into an application, there is little incentive to rewrite its code to take advantage of these enhancements when they eventually appear in the system libraries.

On the positive side, both curses and termcap are no longer specific to the UNIX environment. Third-party suppliers have created versions for MS-DOS and OS/2 on personal computers, and for Digital's proprietary VMS operating system.

'Look and feel' for character terminals

None of the *de facto* standards for character terminal control addresses the issue of 'look and feel'; they simply provide the programmer with a means of displaying characters which is independent of the type of terminal equipment in use.

Figure A2.11 shows that there is no style guide to tell the programmer how to lay out data on the screen, and how to handle user interaction. Similarly, there is no standard high-level toolkit which could encourage programmers to adopt a particular style by making it easy to handle layout and interaction in a particular manner. Consequently, application developers must create their own toolkits and style by building on the lower-level libraries. The result is that character-based applications which use curses, terminfo, or termcap do not have a common 'look and feel'.

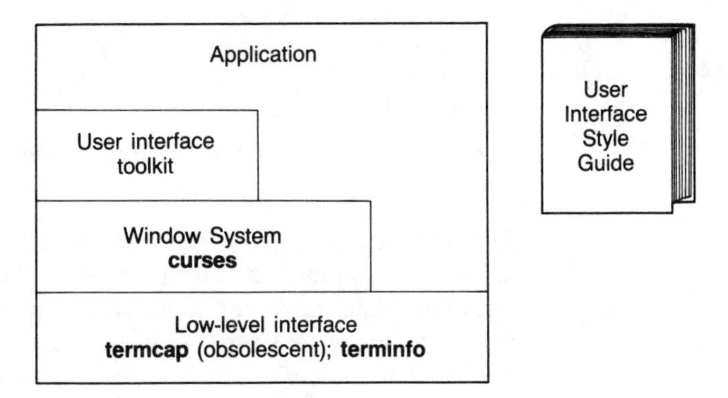

Figure A2.11 User interface model and terminal libraries

Several proprietary libraries have been developed to fill the need for a user interface toolkit, but none has come close to achieving *de facto* standard status. AT&T's System V, Release 4 of the UNIX operating system includes additional libraries which build on curses, and allow applications to construct forms, panels (windows) and menus on character terminal screens. This extended terminal interface (ETI) is defined in the SVID, but does not mandate a particular 'look and feel' for applications which use it.

AT&T has published a *Character User Interface Style Guide*, but this applies mainly to the use of another subsystem, the Forms and Menu Language Interpreter (FMLI). FMLI is not a library, but an application designed to impose a 'look and feel' on underlying applications. In this way it allows a 'look and feel' to be built into character-based applications, rather than pasted on top. It is, however, not defined in the SVID and therefore its usefulness must be questioned.

Application generators as a source of 'look and feel'

A number of application generators and fourth-generation languages impose a 'look and feel' on the developers using them. They do this by constraining the freedom of the programmer, often by 'hiding' low-level mechanisms such as curses and terminfo. As a result, all applications developed using a particular generator tend to share a common 'look and feel' even if no explicit style guide accompanies the development tool (see Figure A2.12). This property has discouraged widespread use or acceptance of any one applications generator. Software vendors have judged it important that their own applications should not look similar to those of their competitors, and so have been resistant to using tools which impose a 'look and feel'.

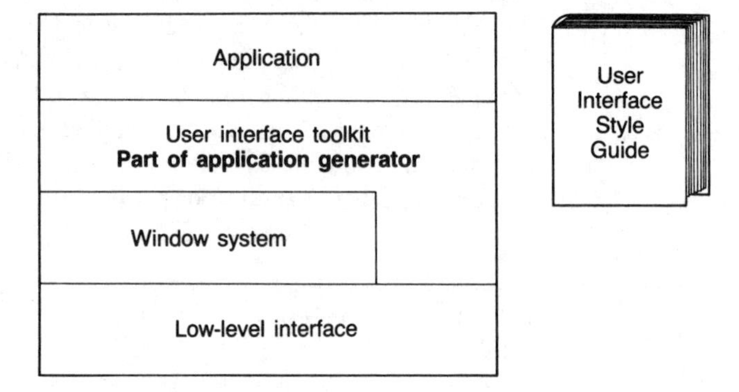

Figure A2.12 User interface model and application generator

Considerations of product differentiation through interface style are less important to authors addressing niche markets, and to the in-house programming departments of large organizations. Consequently application generators are widely used in these areas. However, as with proprietary user interface libraries for character terminals, no one generator has achieved *de facto* standard status through becoming pre-eminent and being imitated.

Summary: character-based user interface standards

- There is no public standard covering the issue of 'look and feel' for character terminals.
- There is no *de facto* standard in this area, nor any pre-eminent proprietary product which might form the basis of a *de facto* standard.
- *Curses*, a windowing library for character terminals is a *de facto* standard which is available for several operating environments, but imposes no par-

ticular 'look and feel' on applications which use it. X/Open has included a definition of curses in its Portability Guide.

- *Terminfo* and its predecessor, *termcap*, are low-level libraries which do no more than make an application independent of the type of terminal with which it is used. Both libraries are *de facto* standards, and are available for several operating environments.
- The facilities provided by AT&T in curses, terminfo and termcap have typically lagged the capabilities of commercially available terminals by several years; facilities which can be considered *de facto* standards have lagged still further. As a result, non-standard extensions to the libraries have proliferated.
- The situation for character terminals is unlikely to change. Development and standardization efforts now focus on more advanced graphical environments where, paradoxically, the situation is considerably cleaner and more simple.

Graphical user interfaces

The addressable element on a character terminal is a character, typically built up of a pattern of dots or picture elements (pixels). Graphical user interfaces (GUI) take advantage of the facilities offered by display screens which allow individual pixels to be addressed. Such screens are known as bit-mapped, because each pixel corresponds to a single bit in memory (or to more than one bit, if the colour or grey shade of a pixel can be specified). Figure A2.13 illustrates this.

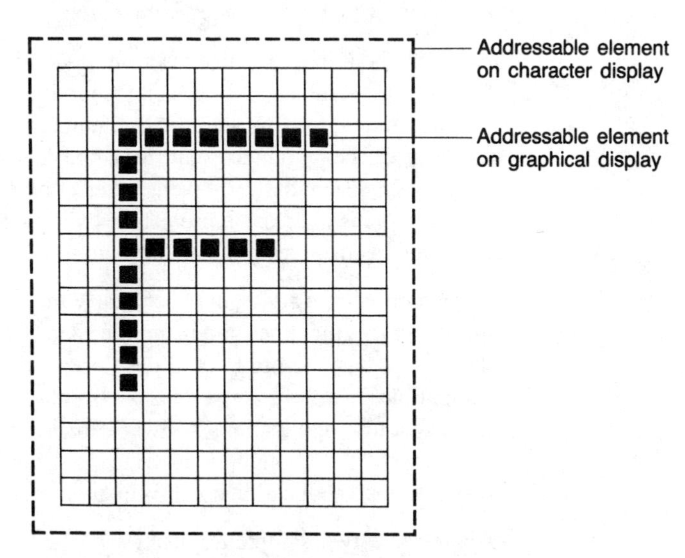

Figure A2.13 Addressable elements on displays

The increased precision of bit-mapped displays allows characters to be presented in a variety of sizes and typefaces, and to be supplemented or completely replaced, with graphical information. This flexibility has a cost. A single character (F), can be specified in just eight bits on most character terminals, but 192 bits are needed on a typical bit-mapped screen. Conceivably it could take 4608 bits to draw a small full-colour picture of the same size. This means that the communication channel between a bit-mapped screen and a computer must be much faster than that for a character terminal if the screen update speed is to appear similar. In fact, all practical methods for the control of bit-mapped screens incorporate short-cuts to allow the rapid display of text without the need to specify the state of each pixel. None the less, the communications and storage overheads are significant.

The designers of user interfaces often take advantage of the graphical capabilities of bit-mapped screens to present a 'desktop' covered with: windows (self-contained rectangular areas); icons (small pictures which indicate system facilities); and menus (vertical or horizontal lists of alternative actions or facilities). The final element in this equation is a pointing device—typically a mouse—which can be used to move a cursor on the screen. Interfaces of this type are sometimes known as WIMPs, from the initials of their four elements windows, icons, menus, pointer.

'Look and feel'

Graphical user interface technology was developed at Xerox' Palo Alto Research Center (PARC) in the 1970s, but was not exploited by Xerox. Instead, others have followed Xerox' lead, and produced systems designed for the commercial market. For several years, Xerox made no attempt to enter formal licensing agreements with these companies, or to collect licensing fees. It seemed that anyone could use Xerox' developments without charge. This situation has changed of late as some companies (most notably AT&T) have actively sought licences from Xerox.

Apple introduced a graphical user interface on its Lisa computer in 1983. The Lisa was a precursor to the Macintosh, which found acceptance as a 'friendlier' alternative to IBM's personal computer, and introduced a wide public to a user interface with consistent 'look and feel' (see Figure A2.14). Apple has gone to great lengths to prevent other companies from replicating its 'look and feel'. This aggressive protection by Apple of its graphical user interface has made licensing from Xerox attractive.

Apple published a style guide very early in the life of the Macintosh. Only by following its guidelines and using the Mac's high-level user interface tool kit (known as the Toolbox) could developers create programs which would be compatible with future releases of the computer's system software. Consequently, applications software for the Macintosh has a very uniform 'look and feel', although the interface is sufficiently flexible to allow features which set one application apart from another.

Apple gave two sample applications free with every early computer sold: MacPaint, a pixel-painting package and MacWrite, a simple 'what you see is what you get' (WYSIWIG) word-processor. By demonstrating the recommended 'look and feel' for the Mac, these influential programs educated users (as to what they should expect), and programmers (as to what they should provide).

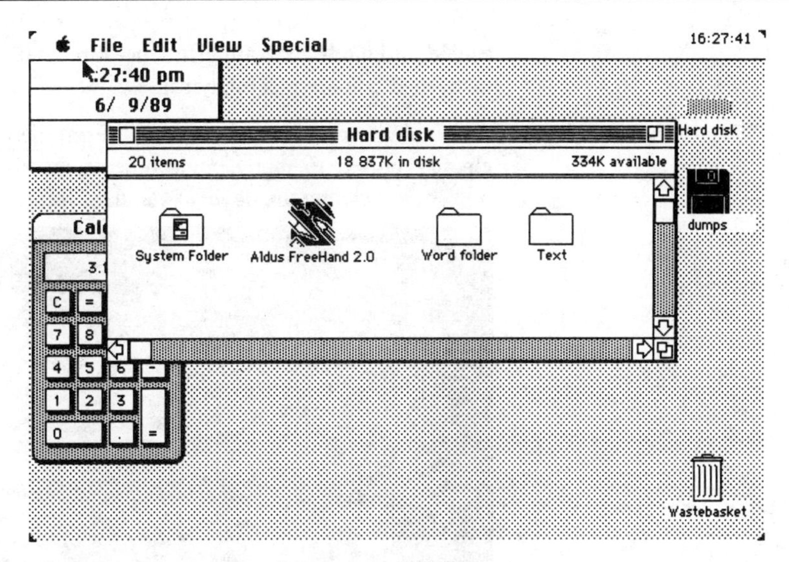

Figure A2.14 Example of Macintosh 'deskTop'

The Macintosh cannot be regarded as an open-systems standard, since Apple protects both the appearance of the system's user interface, and many aspects of the computers which support it. The Macintosh user interface is available only on the Macintosh, and application programs which use it require extensive modification if they are to be moved to other graphical environments. Nevertheless, despite its proprietary nature, the fact remains that the Macintosh serves as an important point of reference for the designers of alternative GUIs.

In 1987, IBM and Microsoft jointly released a GUI for the IBM range of personal computers. Presentation Manager (PM) (see Figure A2.15) is closely tied to IBM's OS/2 operating system, but is also available for MS-DOS in the shape of Microsoft Windows. It will also be incorporated into IBM's Systems Application Architecture (SAA).

Presentation Manager is not yet (1990) available for computers running the UNIX operating system, but OSF Motif, a derivative of PM, is. Motif, illustrated in Figure A2.16, provides a 'look and feel' which is very similar to that

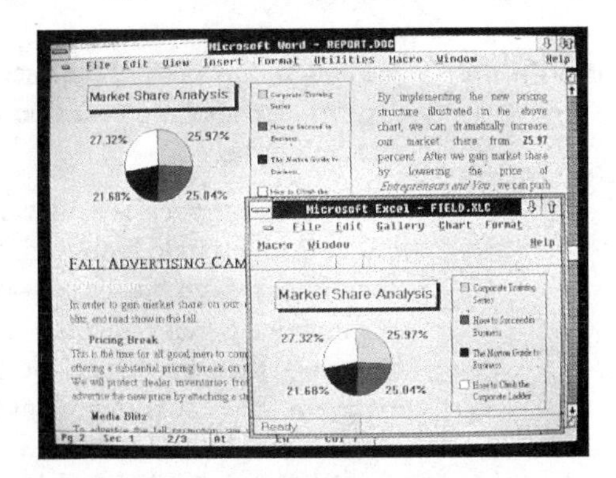

Figure A2.15 Microsoft/IBM Presentation Manager 'DeskTop'

of PM, although there are differences in detail. For example, Motif uses more restrained colours and thinner lines.

Available in source code at low cost from the Open Software Foundation, and already ported to many environments, Motif appears to be well-placed at present to become the *de facto* standard for an open GUI.

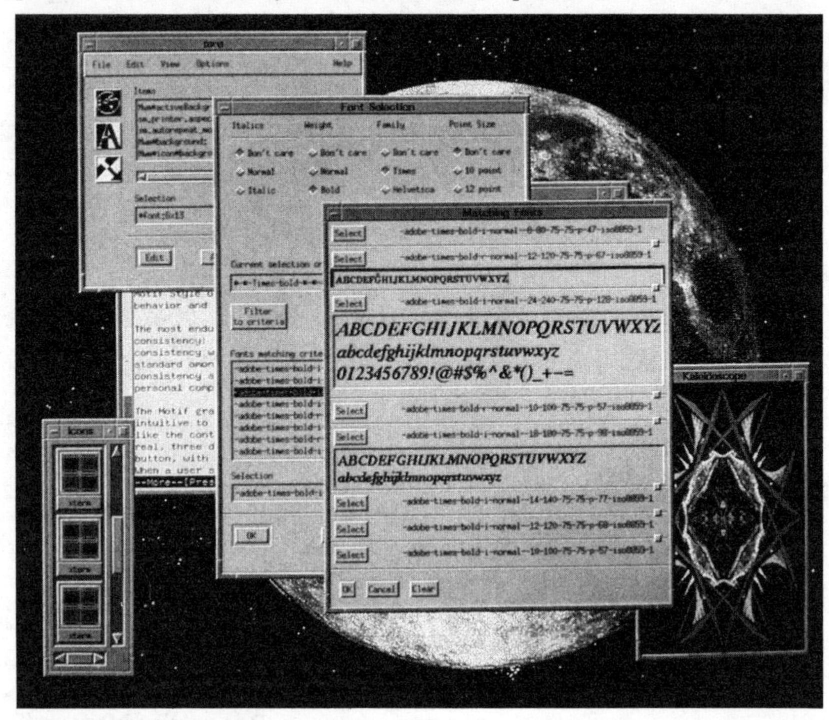

Figure A2.16 OSF Motif 'desktop'

The main commercial competitor to OSF's Motif is Open Look (Figure A2.17) developed by Sun Microsystems and AT&T, with key elements of its technology licensed directly from Xerox. Although it is shipped by AT&T as part of UNIX System V, Release 4 and is available to developers in source form at the same price as Motif, its acceptance has so far been slower.

Technology for windowing systems

Although they present the user with somewhat different interfaces, Motif and Open Look are both based on very similar technologies. Figure A2.18 shows the similarities.

Both user interface schemes closely follow the model introduced in Figure A2.2. Both provide toolkits for use with the X Window System, developed by the X consortium at the Massachusetts Institute of Technology (MIT). Both communicate with display devices using network protocols, rather than requiring that the display be local to the computer running the application program. The most important technical distinction between Motif and Open Look is that Open Look, as well as using the X Window System, can also use Sun Microsystems' Network-extensible Window System (NeWS). The impli-

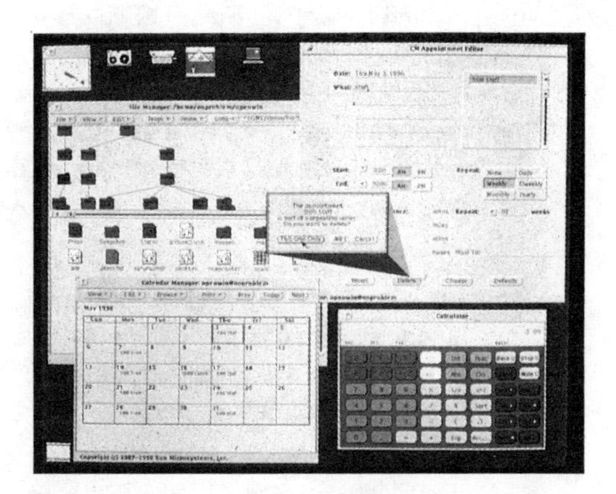

Figure A2.17 Open Look 'desktop'

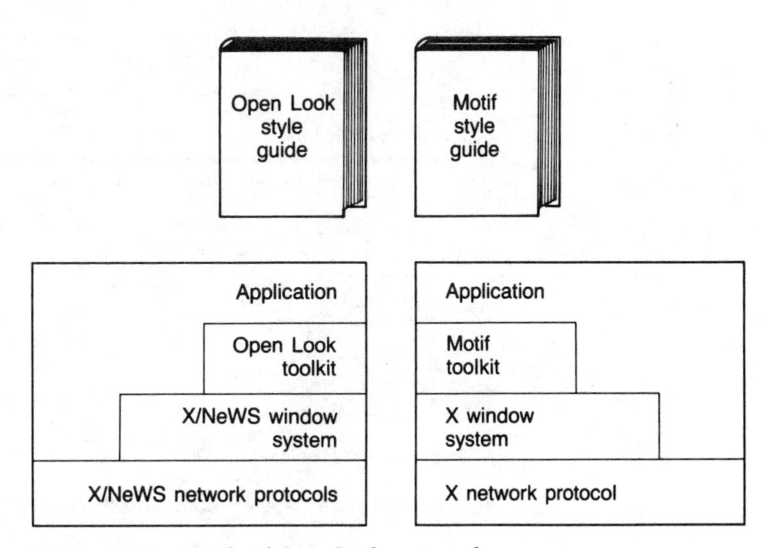

Figure A2.18 Motif and Open Look compared

cations of this difference are discussed below, but first it is useful to introduce another idea, common to both approaches.

Client-server windows architectures

The use of a network allows a windowing system to adopt a *client-server* architecture. The clients are application programs which require access to screen windows and to user input from keyboard and pointing device. The servers are system programs which, by controlling user terminal hardware, provide these facilities.

Figure A2.19 illustrates this arrangement in a simple system with two display devices, each with their own server programs (A and B), and two client programs (1 and 2). Client program 1 is running on a computer which has no display facilities of its own. By accessing server programs A and B over a local-area network, client 1 can simultaneously display windows on the display devices managed by those servers. Client program 2 runs on a workstation computer which has local display facilities. By communicating with the local server, client 2 can control a window on the workstation's screen. At the same time, it can access server A over the network, and so display a window on a remote terminal.

Because use of the network is designed to be transparent, client programs have no knowledge of the location of server programs, The same communications protocols are used whether the server is running on the same machine as the client, or on a remote system.

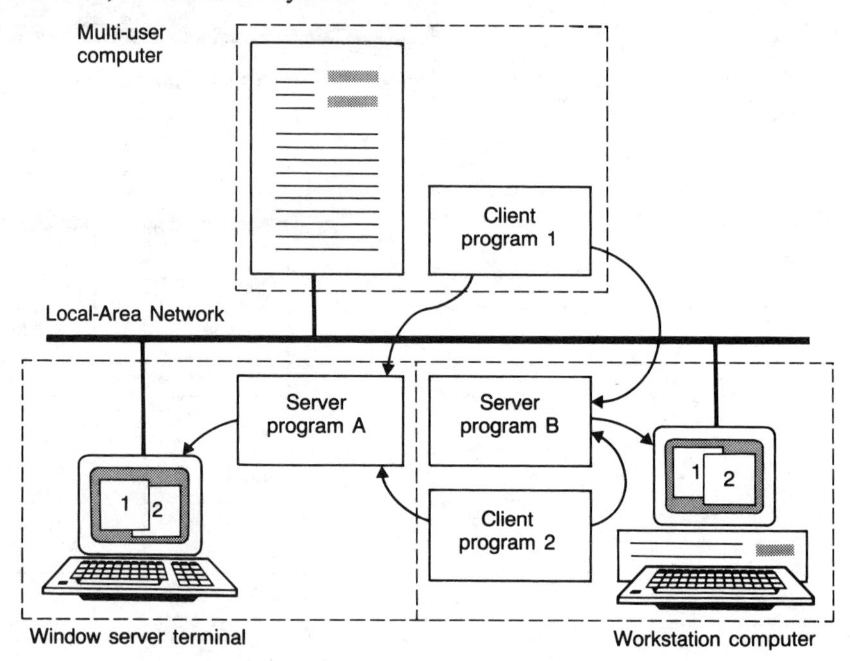

Figure A2.19 Client-server window architecture

With window systems of this type, the server program runs on the equipment which is on the user's desk, while the client program may be remote. This situation is the reverse of the more familiar situation, where a database server provides a shared information store for a number of personal computers and the client programs run on desktop PCs. Then it is the *server* program which is remote from the users. In either case, the server always controls hardware and consequently must be 'close' to the hardware that it manages. In the case of a window system, the server is close to the screen; for a database system, it is close to the disk. Client programs, which are generally applications, can reside anywhere in a networked system provided that they have access to the resources that they need.

The diagram in Figure A2.19 shows server program A running on a window server terminal. The server is the only program which runs on this device. It is incapable of running applications software. It provides the user with the facilities of a high-quality graphics display, but relies on remote computers to run the applications which use it. This situation is the modern analogue of the character terminals of the past. Character terminals have always required a remote computer to run the applications which drive them.

Figure A2.20 shows the components of a window server terminal. Containing server software in read-only memory, these devices support one type of windowing system. The proliferation of such products indicate that the market is standardizing on a technology known as X Windows.

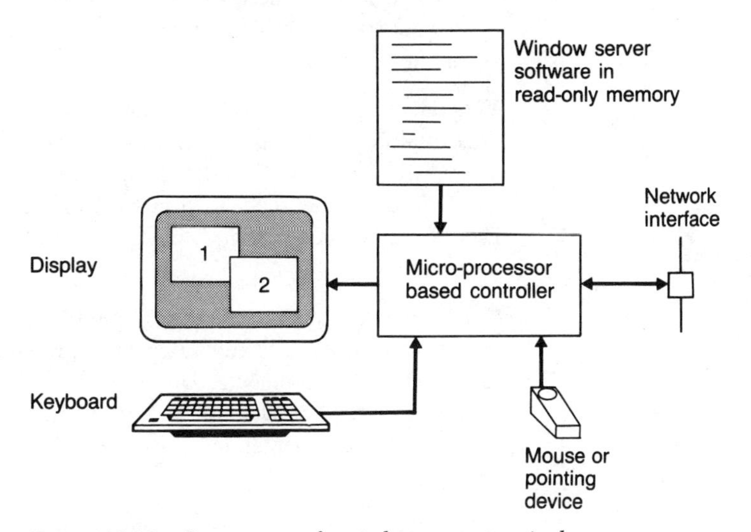

Figure A2.20 Components of a window server terminal

The X Window System

Work started on the X Window System at the Massachusetts Institute of Technology (MIT) in 1984. In the manner which characterizes the naming of many developments associated with the UNIX operating system, X is so called simply because it is a development of an earlier system called W (just as the C language was developed from an earlier language called B).

MIT is very active in developments which will affect the future of computing with networked systems. Its 'Project Athena' is funded mainly by Digital Equipment and IBM and is investigating the use and control of very large, mixed-vendor workstation computers in an academic, yet security-conscious, setting. However, of the many developments associated with Project Athena, only the X Window System (X) has yet reached the mainstream of open systems computing.

X development is backed and funded by the X Window Consortium, a broadly-based, cross-industry grouping of computer suppliers among which are Digital Equipment, Hewlett-Packard and IBM. Obtainable in C language source-code for little more than the cost of duplication, and already shipped in binary form for many different types of computer, X has passed through a number of publicly released revisions, culminating with version 11.3. Future releases are likely to be upward-compatible with 11.3, as its widespread use now precludes incompatible changes.

X, widely accepted as a *de facto* standard, is unusual in that it is the result of a cooperative effort between the computer industry and a leading university, and has been developed in a remarkably short period.

To some extent, the speed of development of X has resulted in problems. For example, the networking aspects of X were not explicitly synchronized with emerging standards for open-systems networking, and the handling of international character sets within X is, in some instances, at variance with international standards. These and other issues are expected to be resolved during the formal standardization process for X, which is now under-way, only five years after the project's inception.

Version 11.3 of the X Window System is a *de facto* standard endorsed by all three major groups associated with the UNIX operating system standards. The Open Software Foundation specifies it in its 'level zero' standards; UNIX International includes it in the 1989 edition of the System V Interface Definition; and X/Open devotes a volume of its Portability Guide (XPG) to the system.

In addition to its widespread availability on implementations of the UNIX operating system, X is supported by a growing number of alternative platforms. Digital Equipment's DECWindows product for its proprietary VMS operating system builds on X, and Microsoft has announced that it will provide an implementation of its Presentation Manager 'look and feel' in terms of X.

A number of X terminals have come on to the market, and software products which allow personal computers to act as X terminals are available. Among these is Apple Computer's product, which allows the Macintosh to act as an X server, and marks the first time that Apple has provided a means for the Macintosh to present a graphical user interface other than its own.

X is now moving rapidly into the public standards arena. The IEEE working group 1201 was formed in April 1989 to standardize two aspects of the system: the X Window Toolkit Application Programming Interface (to become IEEE standard 1201.1), and X Window Recommended Practice (to become 1201.2). The current timetable promises standards in late 1991.

Several other public standards bodies have also expressed an interest in topics related to X. For example, the US Computer & Business Equipment Manufacturers' Association (CBEMA), an accredited standards body, is to develop a 'tiled raster graphics' addendum to the ISO 8613 standard, the international standard for Office Document Architecture (ODA) and Office Document Interchange Format (ODIF).

A possible problem in future upgrade paths for X concerns display resolution. X encodes much information as bit-maps and assumes that displays show somewhere between 70 and 100 pixels, or dots to the inch (dpi). This is acceptable today, but may present problems when higher resolutions become available. For example, unless all the bit-maps were changed, a 300 dpi display would show current bit-maps at one quarter to a third of their intended size. Changing the bit-maps may not be a big problem, provided that all the displays addressed by a particular client have similar resolutions. However, if a single client is to cope with several displays of differing resolutions, problems are likely to arise.

As originally developed, X supplied the two lower levels of the multi-layer model for user interfaces. At the lowest level was the protocol suite used to communicate between client and server. Above that was Xlib, a comprehensive set of low-level tools for defining and painting windows, and for informing application programs of events such as keystrokes and mouse movements. Xlib is currently defined in terms of its interface to C language programs, although work is in progress to define bindings for ADA, C++ and FORTRAN.

It is important to realize that Xlib does not mandate any particular 'look and feel'. The toolkit must constrain the flexibility of Xlib if a particular 'look and feel' is desired. In the terminology of X, such toolkits consist of widgets (the word is derived from 'window' and 'gadget') and intrinsics. Broadly speaking, a widget displays something (a button, perhaps), while an intrinsic controls the communication between the widget and the lower-level library.

A wide variety of toolkits has grown up above Xlib. Many have been contributed free of charge to MIT's source-code distribution. As Figure A2.21 shows some of these widget sets comprise complete user interfaces with 'look and feel' and an associated style guide. Others are less ambitious, or more specialized. Examples are MIT's own X Toolkit (Xt), the Andrew toolkit from Carnegie Mellon University (CMU), and Apollo's Open Dialogue. With the exception of Xt, none can be regarded as a standard.

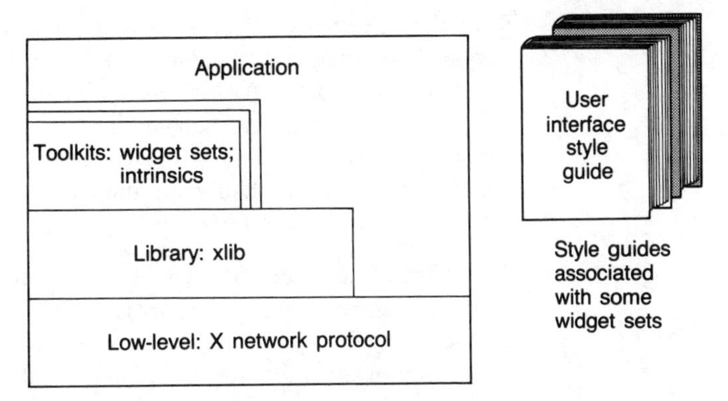

Figure A2.21 Style guides and widget sets

There are, as yet, no formal plans to integrate the separate standardization efforts for X into a single national or international standard. Public standardization will initially concentrate on the lower levels of the system and it will be some time before efforts address standards for widgets.

The Network-extensible Window System

Since shipping its first computer in 1984, Sun Microsystems, a leading vendor of UNIX-based workstation computers, has been committed to windowing technology. Sun's Network-extensible Window System (NeWS) was introduced in 1987, and differed from its predecessors in a number of technical and commercial respects:

- It was designed for hardware independence. Following the client-server model introduced in Figure A2.19, it made few assumptions about the characteristics and capabilities of the hardware running either process.
- Rather than requiring the host to send low-level graphical data to the client, it used PostScript, a higher-level language originally developed for the layout of printed pages, to specify the contents of screen windows.
- Sun offered to license NeWS to its competitors at low cost, in the hope that the technology would rapidly become a *de facto* standard for window management. This, Sun hoped, would expand the market for all vendors by short-circuiting the customer resistance brought about by a lack of standards.
- When, in late 1987, Sun formed a technological, and later, commercial, alliance with AT&T, NeWS was put forward as a component of System V, release 4, the UNIX version due for Release by AT&T at the end of 1989.

The use of PostScript is perhaps the most interesting technical aspect of NeWS. The PostScript language allows the specification of complex figures in terms of lines, polygons, area fills, and so on. In this, its capabilities are similar to those of X. Importantly, however, PostScript goes further than X in allowing transformations such as magnification, rotation, reflection and skewing. The characters which make up text are treated as graphical objects, and so can be manipulated in just the same way as any other object, allowing very complex typesetting effects to be achieved. X, on the other hand, includes no transformation capability.

Unlike X, PostScript makes no assumptions about the resolution of the display or printing device. Consequently, NeWS is promoted as being better able to accommodate very high resolution displays when these become available. While PostScript can handle bit-mapped images designed for display at particular resolutions, and is even capable of applying transformations to them, most images are better specified in terms of their fundamental graphical components without reference to the characteristics of the display.

The effect of using PostScript in a window server program is that the client programs can become simpler, because the server, interpreting the PostScript instructions sent by the client, is capable of performing transformations which would otherwise be the responsibility of the client. Conversely, a NeWS server is more complex, and requires more resources than an X server because of the need to interpret the various instructions.

The power of the language generally means that the length of the instructions needed by a NeWS server to specify given contents for a window tends to be shorter than those required for an X server. This becomes important when the network link between the client and the server is slow or congested. Finally, because PostScript was originally developed for printers, it is a simple matter for a NeWS client program to produce an exact printed copy of the contents of a window. All it has to do is direct the PostScript instructions used to paint the window to a PostScript-equipped printer. A client program using X is almost certain to have to generate printing instructions which differ markedly from those used to control windows.

NeWS, then, has some technical advantages over X. But to become a standard, acceptance by hardware vendors, software suppliers and, ultimately, by computer users, is more important. That X now has such acceptance is acknowledged by the inclusion of a combined X/NeWS server in AT&T's System V, Release 4 of the UNIX operating system.

As Figure A2.22 shows, the System V, Release 4 server is capable of handling windows for client programs which use either NeWS or X. However, industry-standard equipment elsewhere on the network is likely to support only the X window system, and so will not be usable with NeWS client applications. In particular, NeWS clients will not be able to use the new generation of low-cost X terminal equipment, requiring relatively expensive workstations to provide a user interface. Faced with the alternative of writing NeWS client applications for a comparatively limited hardware base, and X clients for a very broad base, application developers have generally opted for X.

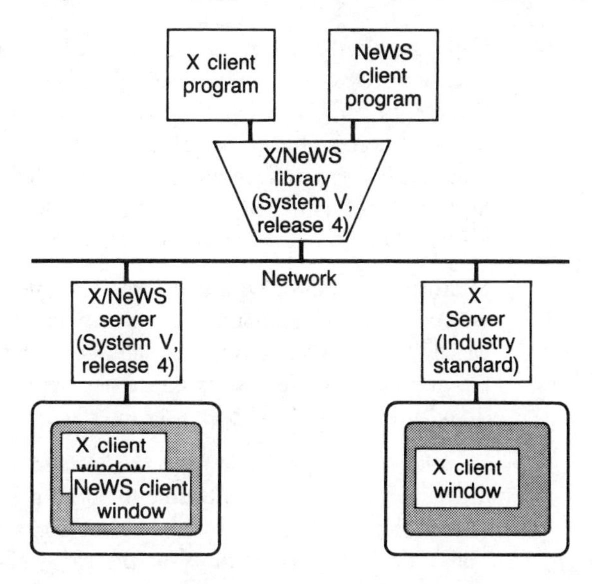

Figure A2.22 System V, Release 4 X/NeWS server

PostScript

PostScript has been mentioned in connection with NeWS. In fact, Sun Microsystems pre-empted Adobe, the developer of PostScript, by marketing NeWS, a PostScript-based windowing system, in 1987. Display PostScript, Adobe's own windowing product, was not released until the following year. It has been adopted by NeXT Inc, and is offered by IBM as an option on its 6000 series of UNIX-based computers.

PostScript is widely used in medium- and high-quality page printers (laser printers, for example), and currently commands a price premium relative to alternative and less powerful technologies. Although, as a page layout language, PostScript has no competitors, its high price and Adobe's proprietary attitude have probably limited its market penetration.

However, if PostScript facilities are integrated with the X Window System, the technology gap between X and NeWS may be closed.

'Look and feel' revisited

Each of the two major groups which are providing products for the open-systems world—the Open Software Foundation (OSF) and Unix International (UI)—has a candidate for the standard user interface for open systems in the form of (respectively) Motif and Open Look. Both take the form of toolkits which use the facilities of the X Window library, and both are available in source-code form at low cost so as to encourage systems implementors to include them in their offerings.

Motif is the result of a request for technology (RFT) made by the OSF in August 1988. After holding an open evaluation of the many products put forward, the OSF decided to produce Motif by combining a user interface from Hewlett-Packard with user interface libraries from Digital Equipment and the X Window System from MIT.

Programming considerations

It is possible that the industry may settle on at least two rather similar user interface styles to cover the spectrum of open systems. These are likely to be OSF/Motif for UNIX-based systems, and Microsoft's Presentation Manager for IBM-compatible personal computers. In addition, Apple's proprietary Macintosh user interface will persist, AT&T's Open Look may capture some market share, and IBM's proprietary systems will use their own variant of Presentation Manager as a component of SAA.

Microsoft has estimated that it takes 1.5 times as long to write an application with a graphical user interface as it does to write one which is character-based (Figure A2.23). Approximately one third of the development time for the graphical application is concerned with the particular libraries and the 'look and feel' of the chosen user interface.

To re-target the application for an alternate 'look and feel'—for example, to move a Macintosh Application to OS/2's Presentation Manager—takes a further third of the development time for a single graphical environment. Thus, the development time for the same graphical application targeted at two different user interfaces is double that required to create a single traditional character-based package.

In other words, the development of graphical applications is going to be expensive, and each additional user interface style needing to be addressed will make it even more expensive.

It is in the software developers' interests to target as few interface styles as possible. This provides a major force for the adoption of a single standard. As applications written to a particular style become available, users could accel-

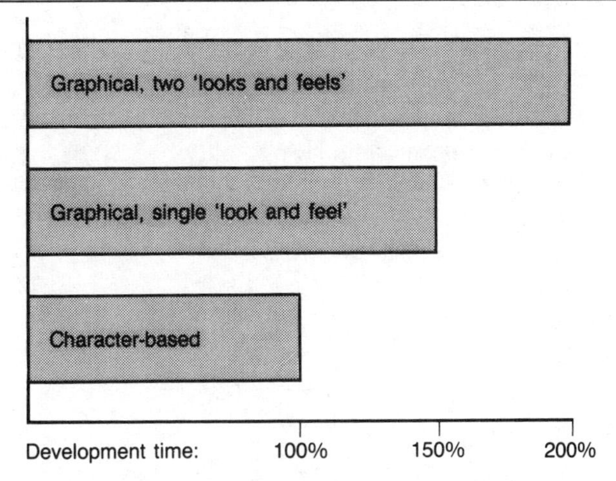

Figure A2.23 Comparison of development times for character-based and graphical applications

erate the standarization process by demanding that further packages adhere to the interface chosen by early developers.

While Motif has a similar 'look and feel' to Presentation Manager (PM), the programming interfaces and underlying structure of the two packages differ considerably. A program written for Motif will require considerable change before it can be used with PM—although not as much as a Macintosh program. This presents developers wanting to target both Motif and PM with conversion costs that they would prefer to avoid. Microsoft has announced its intention to provide an implementation of PM which, like Motif, will ride on top of the X Window System. This would allow applications written to a single user interface—PM—to run without modification either on OS/2 or on open systems supporting X.

Summary: graphical user interface standards

- There is as yet no public standard for graphical user interfaces.
- OSF/Motif could become a *de facto* standard for a uniform 'look and feel' for open systems. Many systems vendors have implemented on their systems and many applications developers have expressed their intention to use it for their future products.
- Other possible contenders for *de facto* standards in this area are: Open Look from Unix International; Presentation Manager for UNIX, from Microsoft; NextStep from Next.
- X Windows, from MIT, is a *de facto* standard for windowing systems, and will undoubtedly become a public standard. It is supported by all constituents of the computer industry.
- Sun's NeWS, which forms an optional part of UNIX System V, release 4, and can be freely licensed from Sun Microsystems, seeks to become a *de facto* standard. However, despite being technically superior to X in some respects, it has not yet been widely adopted.

- The inclusion in UNIX System V Release 4 of a combined server for X and NeWS is a tacit admission that most application software vendors are writing for the X interface. There is no public standardization effort in prospect for NeWS.
- Although it appears that several standards are emerging for the graphical user interface, costs of writing software to more than one will increase development costs significantly.

Appendix 3 Internationalization

In producing applications for use in an international marketplace, there is the fairly obvious problem of the need to translate from one language to another. Language itself is not the only problem. For example, accounting conventions and taxes vary from country to country. Thus, a business software package written for the US must be modified for use in Great Britain or Australia, even though the three countries share the English language. On an apparently trivial level, the English language—and consequently, the 'Latin' character sets used by American computers—lacks the accents needed in many European languages. Europeans, however, fare better than the Japanese and other Asian and Pacific groups, which do not use Latin characters.

To deal with these issues, computer environments and software packages targeted at particular cultures have emerged all around the world. The Japanese write software for Japanese users; the Swedish for Swedish users, and so on. The price that the software commands tends to vary according to the size of the target market. An accounting package which communicates in Swedish and understands Swedish tax laws is likely to be considerably more expensive than a similar package targeted at the US or Great Britain, where the market is much larger.

Until recently, internationalization has not been considered a problem for operating system software, since operating systems tend to use few national characteristics. Those which are embedded in the operating system—the format of the date or time, for example—can be hidden or changed in particular markets through programming. In any case, the characteristics of an operating system should not be seen by most users of computers; trained operators or systems programmers should be able to protect other users from any nationalistic traits. As a consequence, the world market for operating system software is currently dominated by products developed in the US and customized for other markets.

Application software is a different matter. The need to communicate with users in their native language can easily tie a package to its own home market. While many American software packages have been successful around the world, their market is limited to the number of users prepared to use them in the English language. Software authors have recognized this, and have created packages which can be customized to meet the needs of particular cultures. Some do this customization themselves, or through subsidiaries; others appoint agents to make the changes, and then sell the product in a particular market.

There are two problems with this approach. First, the underlying operating system may not be sufficiently flexible to support the needs of a particular culture. As an example, until very recently most implementations of the UNIX operating system (or, more strictly, of its system utility programs) were unable to process character sets sufficiently rich to include all Western European accented characters—and certainly not Arabic, Cyrillic, Hebrew or Japanese scripts. Secondly, in the absence of standardization, there is no commonality between one software supplier's customization and another's.

Although specialists in translation exist, knowledge of one customization procedure is of little use when it comes to customizing the next software package. A new learning process is involved each time, pushing up costs—and consequently, the price to the ultimate user. System administrators also have problems if they must maintain a number of software packages which have been customized in widely differing manners. In an effort to bring standards to this arena, increasing emphasis on internationalization is being given within the standards groups.

Strictly, internationalization is the first half of a two-part process. An internationalized operating system or application is independent of any particular culture (or *locale*, in the terminology of the field), but provides the facilities for all. Subsequent localization adds a particular set of characteristics, making it suitable for a particular locale. A second localization accommodates a second locale, and so on.

The designers of systems which accommodate internationalization must not think that national boundaries equate to locale boundaries. As Switzerland's four official languages show (Figure A3.1), this is not the case. Any comprehensive system must allow each element of a locale description to vary independently.

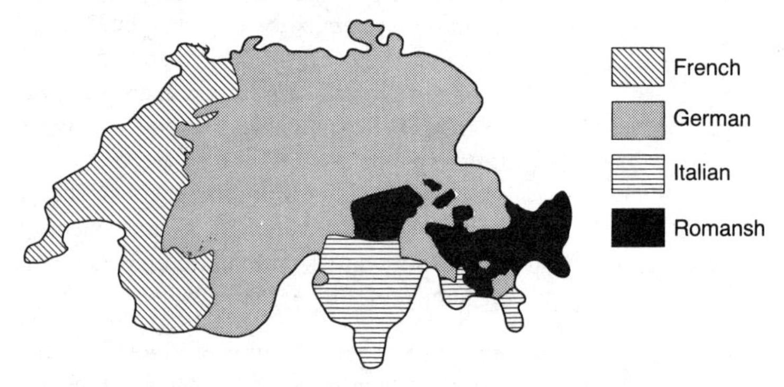

Figure A3.1 Switzerland shows locale and country not equivalent

The user view

Many users of computers are not aware that internationalization is an important issue for them. If software exists to fill their particular needs—including the need for an understanding of their own cultural conventions—they are satisfied. The possibility that the same software might also satisfy the needs of a different culture is not a consideration. Indeed, if adding the features needed to support such customization makes a package larger or slower, its less flexible competitors may be seen as more attractive. However, if the software developer incurs costs as a result of having to import several versions of a product for different languages, those costs will be passed on to all the users of the product.

Multinational companies such as British Airways or DHL, and government bodies such as the European Community (EC) or the North Atlantic Treaty Organization (NATO), have obvious requirements for internationalized products, and are increasingly demanding them. The ability to specify a single application package for use across many cultures is seen as a means of building corporate unity and of promoting internal communication, as well as a path to cutting the costs of support of many different products for each territory.

Software suppliers are also beginning to see the merits of standards for internationalization. A home market may once have been considered sufficient for a software product to recoup its development costs and move into profit, with overseas sales coming as a welcome bonus, but this is no longer the case. Spiralling development costs mean that many classes of software must now be sold internationally in order to reach the sales volumes necessary for profit. Consequently, it makes sense to write software which can easily be customized for the global market.

Steps towards internationalization

Before attention was paid to internationalization, those parts of an application program which had cultural dependencies tended to be dispensed throughout its code. Anyone wanting to adapt the program to the needs and expectations of another culture usually had to search through every module to identify the relevant sequences, and replace them. Figure A3.2 illustrates this process.

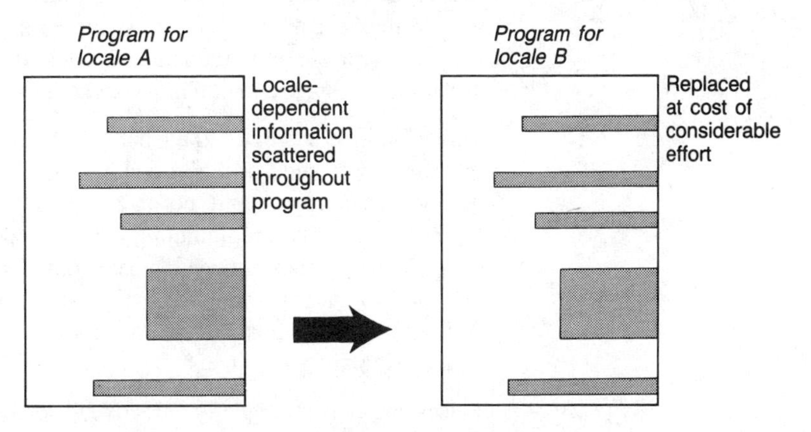

Figure A3.2 Adapting traditional program to new locale

In some cases, this could be simple. For example, error and prompt messages can be translated into another language, provided that the translated version still fits in the space available in memory or on the screen. In other cases, such as changes required in the display of a date or a number, the program code might itself require modification, with a consequent risk of introducing new bugs in the software. Often, the additional programming could be extensive and complex, requiring a high level of programming skills on the part of the translator, in addition to a knowledge of the target culture.

Application of modular programming techniques could confine culture-dependent messages and programming to a single software module (see Figure A3.3). The culture-independent part of the program then references this module whenever a culture-specific service is required. This eliminates the tedious and error-prone process of searching the entire program prior to every translation, but has little effect on the rest of the job.

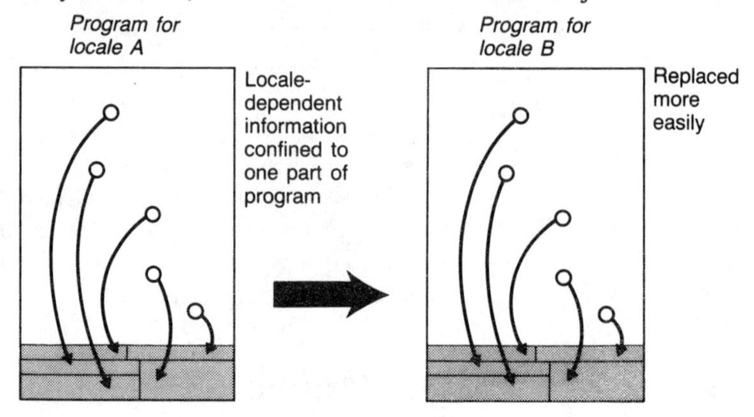

Figure A3.3 Adaptation of program with localized locale-dependent information

A more flexible technique, shown in Figure A3.4, holds the culture-dependent information in a specialized data file external to the program. This makes the program more complex. It must contain code to retrieve information from the data file, and must accommodate all foreseen ways of, for example, representing numbers and dates. It must choose a particular representation in response to the information that it finds in the data file.

This technique has two big advantages. First, localization no longer requires that the source code of the program is modified. The process can be carried out by translators who are not in possession of the source code and who do not need extensive programming skills. Second, because the source code is not modified, it does not need to be recompiled to a binary program following the conversion of the application. A single binary version of the program suffices for all markets—or, at least, for all those markets with needs foreseen by the original programmers.

This property is particularly attractive to software developers, as it means that they can retain the source program, rather than having to pass it to translators who may compound maintenance problems by modifying the code. The problem of maintaining multiple versions of a package for different markets is, in general, more difficult than that of maintaining it for a number of different hardware platforms in the same market.

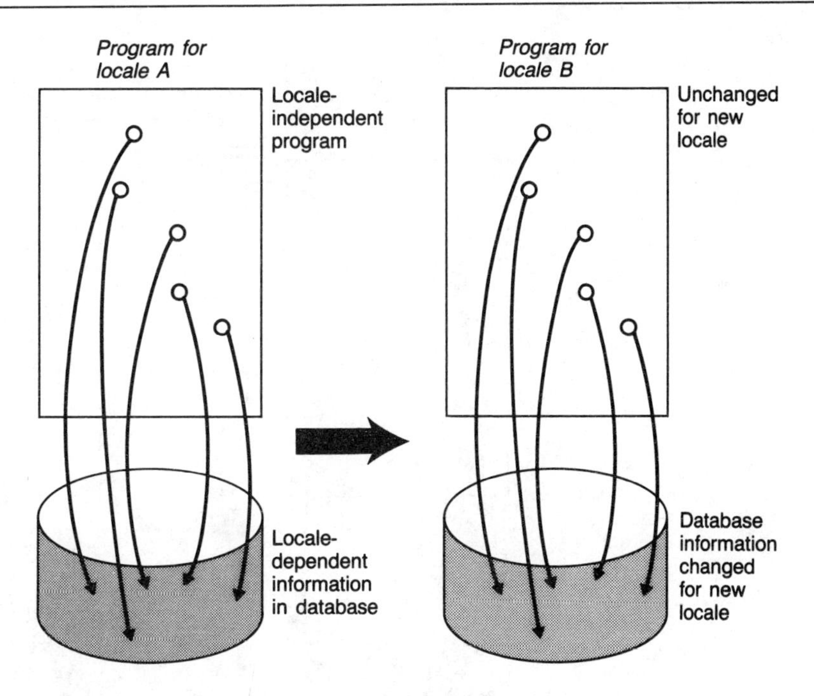

Figure A3.4 Adaptation of data-based program

The schemes shown so far effectively tie an application to a single locale. Once the program or its data file has been modified to address the needs of a particular culture, it loses the ability to work for its original locale. Figure A3.5 shows how this limitation can be eliminated by taking the data file process one stage further, and storing entries for several locales in a single database. The application can then choose between the locales available whenever it is run.

The information about the locale preferred by a particular user is made available to the application by the operating system, usually through *environment variables*. These are user-specific information maintained by the system and set when the system is started, or when the user 'logs-in' to the computer. A system manager, or the person who sets up a particular application package, can set up the environment for each user.

This scheme allows the application to support multiple locales at the same time. Simultaneous users of a single package on a single machine can select different locales. Thus, a multi-user word-processing system could support concurrent use by English, French and German characters, and even provide separate spelling checkers for each language.

There is a model adopted by standardization efforts which is a refinement of the multiple-locale database approach. It lays particular emphasis on defining a set of library functions which allows programs to be written in a manner which is independent of locale. Such a program is said to be *internationalized*.

The names of the environment variables which control the actions of the

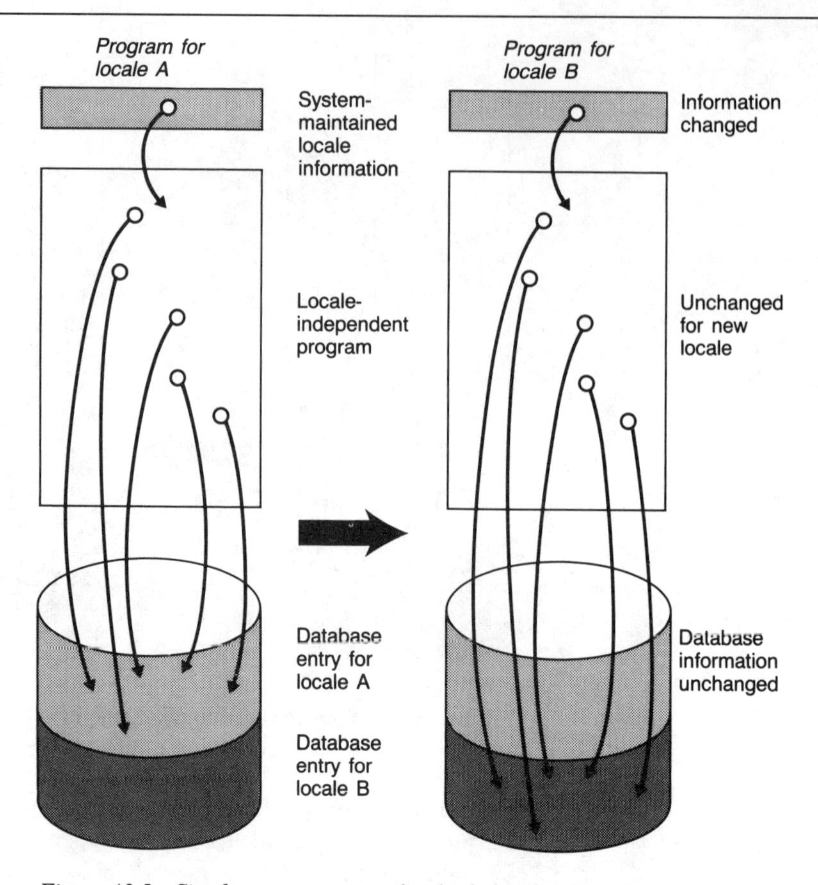

Figure A3.5 Simultaneous support of multiple locales

libraries, and the files in the databases referenced by the routines, are also defined. These databases are not necessarily fully fledged relational databases using SQL—although they could be in a particular implementation. Anything that provides the services required for internationalization is acceptable, and these needs can be satisfied by something much simpler than a relational database manager. (See Figure A3.6.)

Areas affected by internationalization

Character sets

Computers have always minimized the number of bits that they use to represent a character. The most widely used international standard, (ISO 646:1983) allocates seven bits (128 possible characters) to character representations. The 95 printing characters remaining after the allocation of 33 control codes cannot accommodate all the punctuation, digits, letters, accented letters and symbols required even by Western European languages.

As Figure A3.7 shows, fourteen countries manage to define eighteen variants of the basic standard. The same character codes (bit patterns) are used to represent different characters in different countries—or even for different users in the same country.

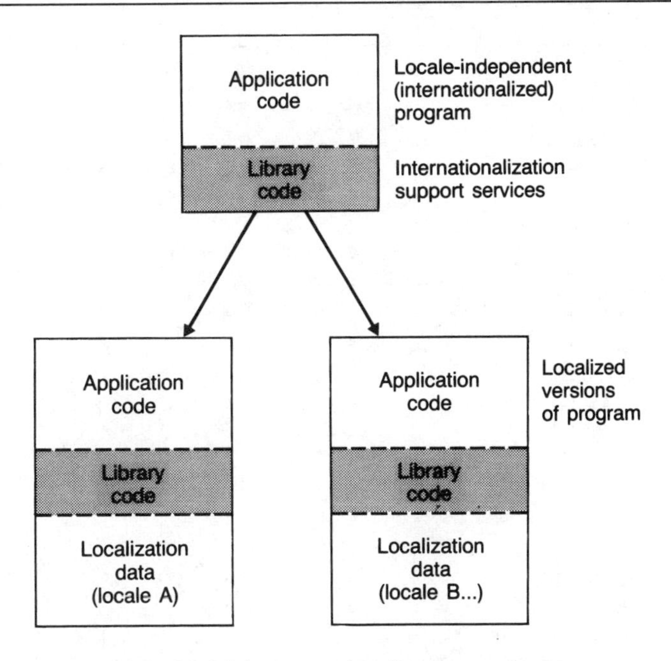

Figure A3.6 Model for internationalization standards

Clearly, a program which can process only seven-bit character sets faces severe problems in coping with multiple locales. Does it handle the code which represents 'Ä' in Germany, but '[' in the USA, as a letter, or as punctuation, for example?

An eight-bit character-set might seem to be the answer. Indeed, a family of eight international standards (ISO 8859 1 to 8859 8) extends ISO 646 to cover all European languages, as well as those requiring Arabic, Cyrillic, Greek and Hebrew alphabets. The 8859 standards extend the 646 standard, and remove the need for national variants.*

The fact that a family of international eight-bit standards is required to address the character sets of only part of the world indicates that the problem is not really solved and that eight bits are not enough. Consequently, work is in progress on (ISO DP 10646) a 'multiple octet' (multi-byte) character set which, by using 24 or 32 bits, will be able to encode every character used in the world.

When this (ISO 10646) standard arrives, it will represent good news for memory and disk manufacturers. Without data compression, data-storage requirements may increase by a factor of four.

The draft standard already defines eight- and sixteen-bit forms of the codes, which are sufficient to accommodate all European and Middle Eastern languages. However, judging from the time taken for the transition from seven to eight bits, transition to such a global standard is likely to be slow and problematic.

*Note that these standards have nothing to do with EBCDIC (Extended Binary-Coded Decimal Interchange Code), the eight-bit standard used on IBM mainframe computers.

<table>
<tr><th>Country</th><th>Standard</th><th></th><th></th><th></th><th></th><th></th><th></th><th></th><th></th><th></th><th></th><th></th><th></th></tr>
<tr><td>ISO</td><td>ISO 646 IR V</td><td>#</td><td>¤</td><td>@</td><td>[</td><td>\</td><td>]</td><td>^</td><td>`</td><td>{</td><td>|</td><td>}</td><td>‾</td></tr>
<tr><td>China</td><td>GB 1988–80</td><td>#</td><td>¥</td><td>@</td><td>[</td><td>\</td><td>]</td><td>^</td><td>`</td><td>{</td><td>|</td><td>}</td><td>‾</td></tr>
<tr><td>Denmark</td><td>DS 2809</td><td>#</td><td>$</td><td>@</td><td>Æ</td><td>Ø</td><td>Å</td><td>^</td><td>`</td><td>æ</td><td>ø</td><td>å</td><td>‾</td></tr>
<tr><td>Finland</td><td></td><td>#</td><td>$</td><td>@</td><td>Ä</td><td>Ö</td><td>Å</td><td>^</td><td>`</td><td>ä</td><td>ö</td><td>å</td><td>‾</td></tr>
<tr><td>France</td><td>NF Z 62–010</td><td>£</td><td>$</td><td>à</td><td>°</td><td>ç</td><td>§</td><td>^</td><td>µ</td><td>é</td><td>ù</td><td>è</td><td>¨</td></tr>
<tr><td>Germany</td><td>DIN 66 003</td><td>#</td><td>$</td><td>§</td><td>Ä</td><td>Ö</td><td>Ü</td><td>^</td><td>`</td><td>ä</td><td>ö</td><td>ü</td><td>ß</td></tr>
<tr><td>Great Britain</td><td>BS 4730</td><td>£</td><td>$</td><td>@</td><td>[</td><td>\</td><td>]</td><td>^</td><td>`</td><td>{</td><td>|</td><td>}</td><td>‾</td></tr>
<tr><td>Hungary</td><td>MSZ 7795/3</td><td>#</td><td>¤</td><td>@</td><td>Á</td><td>É</td><td>Ö</td><td>^</td><td>á</td><td>é</td><td>ö</td><td>ü</td><td>"</td></tr>
<tr><td>Italy</td><td></td><td>£</td><td>$</td><td>§</td><td>°</td><td>ç</td><td>é</td><td>^</td><td>ù</td><td>à</td><td>ò</td><td>è</td><td>ì</td></tr>
<tr><td>Japan</td><td>JIS C 6299</td><td>#</td><td>$</td><td>@</td><td>[</td><td>¥</td><td>]</td><td>^</td><td>`</td><td>{</td><td>|</td><td>}</td><td>‾</td></tr>
<tr><td>Norway</td><td>NS 4551-1</td><td>#</td><td>$</td><td>@</td><td>Æ</td><td>Ø</td><td>Å</td><td>^</td><td>`</td><td>æ</td><td>ø</td><td>å</td><td>‾</td></tr>
<tr><td></td><td>NS 4551-2</td><td>§</td><td>$</td><td>@</td><td>Æ</td><td>Ø</td><td>Å</td><td>^</td><td>`</td><td>æ</td><td>ø</td><td>å</td><td>‾</td></tr>
<tr><td>Portugal</td><td></td><td>#</td><td>$</td><td>§</td><td>Å</td><td>Ç</td><td>Õ</td><td>^</td><td>`</td><td>ã</td><td>ç</td><td>õ</td><td>°</td></tr>
<tr><td></td><td></td><td>#</td><td>$</td><td>`</td><td>Å</td><td>Ç</td><td>Õ</td><td>^</td><td>`</td><td>ã</td><td>ç</td><td>õ</td><td>‾</td></tr>
<tr><td>Spain</td><td></td><td>£</td><td>$</td><td>§</td><td>¡</td><td>Ñ</td><td>¿</td><td>^</td><td>`</td><td>°</td><td>ñ</td><td>ç</td><td>~</td></tr>
<tr><td></td><td></td><td>#</td><td>$</td><td>°</td><td>¡</td><td>Ñ</td><td>Ç</td><td>¿</td><td>`</td><td>´</td><td>ñ</td><td>ç</td><td>¨</td></tr>
<tr><td>Sweden</td><td>SEN 850200 B</td><td>#</td><td>¤</td><td>@</td><td>Ä</td><td>Ö</td><td>Å</td><td>Ü</td><td>`</td><td>ä</td><td>ö</td><td>å</td><td>‾</td></tr>
<tr><td></td><td>SEN 850200 C</td><td>#</td><td>¤</td><td>É</td><td>Ä</td><td>Ö</td><td>Å</td><td>Ü</td><td>é</td><td>ä</td><td>ö</td><td>å</td><td>ü</td></tr>
<tr><td>USA</td><td>X3.4–1968</td><td>#</td><td>$</td><td>@</td><td>[</td><td>\</td><td>]</td><td>^</td><td>`</td><td>{</td><td>|</td><td>}</td><td>‾</td></tr>
</table>

Figure A3.7 National variants of ISO 646 character set
Source: Keitzer *et al.*, *An Extension to the Troff Character Set for Europe*, European UNIX systems User Group Newsletter, Vol. 9, no. 2.

Users of seven-bit codes have compounded the difficulty of transition to international eight-bit standards by inventing their own non-standard eight-bit codes based on seven-bit standards. Common examples are the IBM PC, the Apple Macintosh, and Hewlett-Packard terminals. These devices can represent all characters commonly used in Western Europe, but can communicate with each other only if they confine themselves to the seven-bit ASCII character set. The same consideration applies to communication with equipment which complies to the ISO 8859 standard. In short, a lack of conformance to standards in the past complicates transition to the standards of today.

A further problem for the standards of the future arises because popular computer languages lack facilities for handling character codes which have more than eight bits. The C language, as defined by the ANSI (X3.159) standard, attempts to meet this need, but its facilities are generally judged inadequate, and it will need improvement before an international standard for the language can be agreed. Until such a standard exists, programmers are likely to aviod the facilities, and consequently will limit their programs to the use of eight-bit characters.

Realizing that code set problems will exist for a considerable time, two recent *de facto* standards, PostScript and the X Window System, have approached the problem in a novel way. Instead of requiring particular extensions to the ASCII character set, they define displayable *glyphs* (human-readable representations of characters) by reference to their names—for example 'ampersand' or 'equals sign'. PostScript allows any character code to be associated with any name, removing dependence on a particular code set.

X Windows defines its own 29 bit (32 bits but with 3 pre-assigned) code set based initially on the ISO eight-bit standards (8859-1 to 8859-4). This gives X Windows full coverage of Western European alphabets while leaving ample room for expansion.

With both X Windows and PostScript, existing data using older code sets can easily be translated into the format required by the newer systems. Such translations—and their attendant costs in processing power—are likely to become widespread over the next few years. The requirement for communications between systems using differing code sets is growing rapidly, yet convergence on a common code set takes place only slowly.

Collating sequence

On the face of it, the issue of collating (sorting character-based data) appears to be related to the issue of character codes. This is a misconception which arises because all common character sets are designed to sort non-accented alphabetic data using a simple character-by-character comparison. Real life is not that simple. To give some examples:

- Capitalization is insignificant in dictionary-order sorts;
- 'St.' sorts into the same place as 'Saint' in gazetteers and telephone directories in a number of English speaking countries;
- 'ij' sorts in the same place as 'y' in Dutch;
- 'ab' sorts after 'à' but before 'àb' in French.

In the past, such issues were handled by special-purpose sorting programs written to accommodate particular applications. Such programs have been necessary even where the underlying system appears to provide sorting facilities. Database collation is almost always determined by simple comparison of character codes, and so is unsuitable for many purposes.

Over recent years came the realization that the need for sorting programs could be reduced, and the remaining programs simplified, through the provision of library routines which can perform comparisons based not simply on character codes, but against a range of criteria which mirror the needs of the real world. Work has been proceeding on defining these routines, and on the format of the data tables which drive them, but the results have yet to see wide distribution. Consequently, the facilities are not widely used by practical applications programs.

String searches

Searching for particular words in text is a common function in editing and word-processing software. The UNIX operating system gave rise to a powerful method of specifying the words to be found, known as *regular expression matching*. Briefly, this allows the use of '*wild card*' characters to extend and control matching capabilities far beyond the simple case of an exact match. For example:

- The regular expression 'abc' matches 'abc' anywhere in a line of text;
- '^abc$' matches 'abc' only if it appears on a separate line;
- 'a.c' matches 'abc', 'a!c', 'acc' . . . but not 'ac' or 'abbc';

- 'a.*c' matches 'abc', 'a!c', 'acc' . . . as well as 'ac', 'abbc', etc.;
- 'a[bc]c' matches only 'abc' and 'acc';
- 'a[a-c]c' matches only 'aac', 'abc' and 'acc';
- 'a[^b]c' matches 'aac', 'acc' . . . but not 'abc'.

While regular expressions may not be of interest to many users of text-processing software, their power is attractive to programmers for use during software development, and from within programs.

In standardizing UNIX, the IEEE POSIX working group elected to standardize regular expressions. In order to do this, dependencies on the US ASCII character set had to be removed, so making regular expressions suitable for use in searching texts written in languages other than English. After much work (and debate), the following additional possibilities are now included for example:

- '[[=a=]]bc' matches 'abc', 'ábc', 'äbc' . . .
- '[[:lower:]]bc' matches `abc', 'bbc' . . . but not 'Abc', '!bc' . . .
- '[[=saint=]] bc' matches 'saint bc', 'St. bc' and 'Saint bc', provided that the table which drives the library function contains information about synonyms for 'saint'.

As with collating sequences, practical implementations are only just reaching the market, and application software has yet to make full use of the new facilities.

Date and time

The representation of the date and time varies between locales. Figure A3.8 shows examples, based on literature from Apple Computer.

Again, life is not quite as simple as the figure suggests. Date and time formats vary among different users according to taste and convention, even within one country. This means that it is not acceptable for a system supplier to dictate a single, fixed representation for a particular territory. Any mechanism which varies the date and time formats which are output by a system according to locale should ideally allow fine-tuning by organizations, system managers, users, and application programs.

The Apple Macintosh, which has supported internationalization since its introduction, has such a feature, as do MS-DOS, OS/2, and the 1003.1 POSIX standard for the UNIX operating system. Of these, only the POSIX system copes with the concept of time zones and daylight-savings time, a facility which is important when systems are linked through international networks.

It is worth noting that these facilities provide only for the display of the date and time, not for the capture of date and time information from the user. Data input requires specific programming in each application, but can be internationalized by reference to the information which the system can provide about the locale.

Numeric formats

As with the date and time, details of numeric representation vary between locales, particularly in connection with business documents. Figure A3.9 shows some examples.

	United States	*Great Britain*	*Italy*	*Germany*	*France*
Time	9:05 AM	09:05	9:05	9.05 Uhr	9:05
	11:30 AM	11:30	11:30	11.30 Uhr	11:30
	11:20 PM	23:20	23:20	23.20 Uhr	23:20
	11:20:09 PM	23:20:09	23:20:09	23.20.09 Uhr	23.20.09
	5:45 AM EST	10:45 GMT	11:45	11.45 MEZ	11:45
	6:45 AM EDT	11:45 BST	12:45	12.45	12:45
Short date	12/22/85	22/12/1985	22-12-1985	22.12.1985	22.12.85
	2/1/85	01/02/1985	1-02-1985	1.02.1985	1.02.85
Long date	United States		Wednesday, February 1, 1985		
			Wed, Feb 1, 1985		
	Great Britain		Wednesday, February 1, 1985		
			Wed, Feb 1, 1985		
	Italy		Mercoledi 1 Febbraio 1985		
			Mer 1 Feb 1985		
	Germany		Mittwoch, 1. Februar 1985		
			Mit, 1. Feb 1985		
	France		Mercredi 1 fevrier 1985		
			Mer 1 fev 1985		

Figure A3.8 Examples of date and time representation
Source: Inside Macintosh, vol. 1 (with amendments).

	United States	*Great Britain*	*Italy*	*Germany*	*France*
Numbers	1,234.56	1,234.56	1.234,56	1.235,56	1 234.56
Currency	$0.23	£0.23	L. 0,23	0,23 DM	0,23 F
	($0.45)	(£0.45)	L. -0,45	-0,45 DM	-0,45 F
	$345.00	£345	L. 345	325,00 DM	325 F

Figure A3.9 Examples of numeric representation
Source: Inside Macintosh, vol. 1.

Again, MS-DOS, OS/2 and POSIX (as well as the Macintosh) provide facilities which allow a program to vary the manner in which numbers are displayed according to the locale.

Messages

For each application, and for the operating system itself, error messages, user prompts, and allowable values for user responses vary according to locale. An internationalized program should not itself contain the character strings used for these functions, but instead should retrieve them from a database keyed by the current locale. An identifier for the program, and an identifier for each message is then used by the application program.

X/Open introduced a definition for such a message catalogue system in the second edition of its Portability Guide, carrying the description into the third edition without major change. Practical implementations of the system are slowly becoming common. However, it is unlikely that applications software developers will use the technique until it is provided across a broad range of system types.

Standards for internationalization

Uniforum, through its Technical Committee, has for many years led much of the standardization work on internationalization. X/Open has also been consistently a leading contributor to the work. Among individual computer suppliers, Hewlett-Packard's early contribution stands out, although many others have also played a significant part.

A problem for those concerned with internationalization is where best to make their contribution. As a discipline, internationalization affects the formulation of many standards, yet it is not itself a clearly defined standards area. Rather, it requires a broadening of outlook on the part of all those who create standards for information technology. Consequently, although Uniforum, X/Open and others have worked hard to ensure that emerging standards for POSIX and the C language support internationalization, there is no working group in an accredited standards organization concerned with the topic. Instead, it falls to Uniforum to coordinate activity in an informal manner.

As with many aspects of UNIX standardization, Uniforum had done the early work on the topic in the mid-1980s, expecting to pass the activity to a more formal body when the time was right. Because of the general nature of internationalization, no public body has yet been prepared to take it over. This situation may be resolved by the formation by the IEEE and ISO of *rapporteur groups* to coordinate and report on internationalization issues within their respective standardization activities.

Character sets

At the international level, work on character sets is proceeding slowly within ISO, with the long-term aim of reducing the diversity of character sets in use around the world. Because of the heavy investment by users and suppliers of software and hardware in the entry, processing, storage, and display of data represented in a wide variety of formats, this is likely to take a long time.

An international standard, by definition, must accommodate the needs of the whole world, including those of, for example, Asia, which differ markedly from those of the Western countries. This requirement often slows the development process for international standards.

In order that the immediate needs of a large part of the rest of the world can be met faster, developers of other emerging standards are limiting their ambitions for the moment. In particular, both X/Open and the X Window

Consortium have defined standards which meet the needs of those who use Western European alphabets, based on ISO standards. Both organizations are now working to expand their coverage, which currently excludes Eastern European, Middle Eastern, and Asian character sets.

Specific implementations of *de facto* standards already accommodate Japanese characters. Open Look (UI), PostScript (Adobe Systems), and UNIX System V (AT&T) are among them.

Collating sequence, string searches, date and time, numeric formats

The POSIX (1003.1) standard provides a general framework for environment variables which define various aspects of the locale chosen by the user or the system administrator. Collating sequence is one of these aspects, as are all remaining topics except messaging, which is discussed below. In each case, while POSIX provides the framework, it does not define the mechanism that it supports. In all cases except messaging, this is left to the POSIX 1003.2 standard, which is expected to be approved during 1990.

Until 1003.2 appears, Volume 3 of the X/Open Portability Guide (XPG) provides the most definitive description of internationalization features. 1003.2's definitions, when approved, are likely to be very similar to those in the XPG, since X/Open and its members contributed much of the effort required to draft the IEEE document.

Progress of internationalization features into an international, from a US, standard may be problematic. While ISO is committed to the production of an international standard based on 1003.2, the issues addressed by the IEEE are of questionable applicability to Asian languages—are the POSIX mechanisms adequate for the collation of Japanese or Chinese characters, for example? Additionally, certain aspects of other non-European alphabets, such as text direction, are not covered. If ISO decides that these issues must be covered by an international standard, its appearance will be delayed well into the 1990s.

Messages

The X/Open message catalogue system is defined in the XPG, but has yet to be found in any public standard. There is some concern that the mechanisms defined by X/Open may be too limited. The maximum message length is too short to accommodate a full screen of help information, for example. Public standards bodies may decide to define a new system, which may or may not be compatible with the X/Open scheme. In any event, current workloads and timetables of the standards bodies seem to indicate that little effort can be spared to address the issue of messaging in the near future.

Summary: internationalization

- Internationalization is an increasingly important activity in standardization work. Because it impacts the work in most other areas, there is not a formal standards body dealing with it. Instead, Uniforum and X/Open play leading roles.
- The global market dictates that products must be able to deal with interna-

tional differences within a single version of a product. Support considerations mean that internationalization should be done in such a way as to allow maintenance of a single version of the source code.

- Full support of all languages, including those in Asia, is difficult, and therefore slow. Support of those languges which can be dealt with by eight-bit character sets is provided by many suppliers, but not in a standard way. The eight-bit standards are well advanced in the standards bodies, but not yet in universal use.
- Progress of US-developed standards into full international standards is unlikely to be smooth. Issues of Chinese and other eastern languages have yet to be considered in detail.
- New versions on UNIX System V from AT&T are expected to have full Internationalization support.

Appendix 4 Security

Security is of serious concern both to commercial and to military users of computers. Both groups have confidential data which they wish to conceal from people outside their organizations—and from those inside not authorized to see it.

Where information is held only on paper, locks and security procedures can prevent physical access or at least make it clear when a breach has occurred. Computers complicate the issue because the computer has made it possible to examine, destroy, or corrupt data without leaving clues. The new threats to security created by computer technology must be appreciated by those who use them, and should be countered.

However, computers can also monitor and control security. But this monitoring is not without its cost. Secure systems are expensive to develop, and require skilled administration. Secure systems can also absorb computer resources. In a recent incident in the UK, a newly installed secure system used to hold information on serious fraud offences was found to run extremely slowly because of the checks it had to perform before it could allow users access to data.

The formal science of making systems secure is young, but it is maturing quickly because of commercial pressures on suppliers from prospective users of the resulting technology. As in many other fields, development efforts are concentrating on open systems, even though the methods could apply equally to older, proprietary environments. As a result, open systems in general, and the UNIX operating system in particular, look set to gain a lead in security features, so breaking with their past (and not wholly deserved) reputation for insecurity.

The user view

General interest in secure systems is a relatively new phenomenon, fuelled by increasing reliance on computer networks, and by lurid stories about 'hackers', data theft, and computer viruses.

In the past, those who needed security had to be prepared to implement it themselves, or to pay for an expensive, custom-built system. Military and intelligence users obviously had both the requirements and the necessary budgets.

The financial community also had a strong interest. The Society for Worldwide Inter-bank Funds Transfer (SWIFT), has run a highly secure network for a number of years. The few documented cases of 'wire-fraud' against SWIFT tended to exploit hardware breakdowns and administrative loopholes, illustrating that there is far more to security than the software alone.

As more organizations become interested in security, the demand rises for standard products in the area. Software and hardware suppliers are currently investing heavily to provide the products, but few have yet reached the market. Standards bodies are also gearing up to provide standards for the systems but results will not appear until the early 1990s. As is often the case, products will appear ahead of the standards. And, as usually follows, these products will require amendment in the future if they are to conform to the standards which ultimately arrive.

Secure systems were originally developed for national security applications. Here, the costs of physical security, detailed and intrusive administration, and severe restrictions on users' freedom can be accepted as a price which must be paid in highly sensitive applications.

There is a further price—restrictions on users tend to reduce their productivity. Indeed, the ease of access to information, and the ease with which it can be manipulated, are seen as important reasons for the early popularity and success of the UNIX operating system in software and engineering development environments.

Similarly, personal computers have brought individual users unparalleled facilities in the maintenance and manipulation of their own data, but often allow unimpeded access to that information by anybody who can gain access to the machine.

Users would like reassurance that they *can* have the best of both worlds: that open systems can attain specified levels of security, and that they can do so without compromising their productivity and cost benefits.

This presupposes that there is some standard way of specifying security levels. There is, but even this is a comparatively recent development. As this annex shows, derived standards which describe *how* to reach a given security level have yet to appear. Until they do, suppliers may attain a particular level of security in their products by following their own proprietary route. Users need to know how long this negation of the concept of open systems is likely to persist.

Current standards for security

The Orange Book

The US National Computer Security Commission (NCSC) is the driving force behind most recent developments in computer security. Originally known as the Department of Defense Security Center, the NCSC has come to be concerned with security in a broader sense than military alone.

In 1983, the organization published the Department of Defense *Trusted Computer Systems Evaluation Criteria* (TCSEC), which, because of the colour of its cover, became known as the Orange Book. This became a military standard in 1985. Applicable to any environment, and created before the movement towards open systems became a major force in the market, the Orange Book continues to play a central role in the specification of secure computer systems.

Figure A4.1 shows the layers that must be built on top of a computer system in order that it can be described as trusted or secure. The security policy protects the system and the data that it controls from unauthorized access. The accountability policy allows individual users to be identified, and their actions to be traced. Both policies, while implemented using a combination of hardware and software, require documentation, and must be testable in order that the whole system can pass the evaluation criteria.

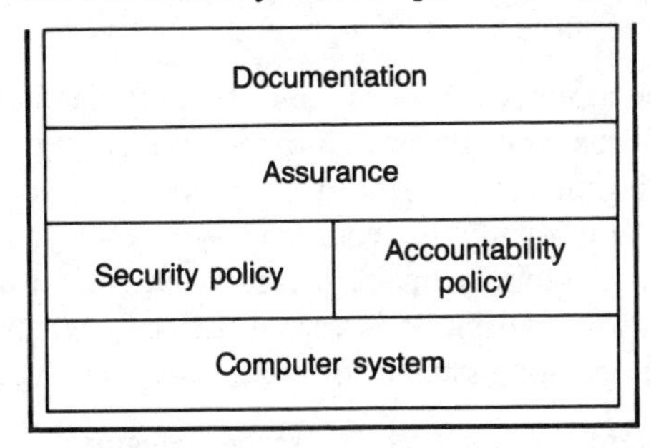

Figure A4.1 Components of a trusted computer system

In the terminology of the Orange Book, a trusted system is one which offers its users and administrators an assurance that a particular, well-defined level of security is attained. In order to achieve this, it is not enough for the computer's hardware and software simply to incorporate security facilities. The facilities must pass a formal evaluation, and their purpose and use must be properly documented.

Without testing, a system may harbour unsuspected loopholes in its security. Without documentation, the facilities that it offers may be misunderstood, misused, or ignored. Just as testing houses have sprung up to check conformance to POSIX operating environment standards, and to OSI communications standards, so there is a growth in specialist companies which evaluate systems against the security requirements of the Orange Book.

Levels of trust

The Orange Book defines four classes of trusted system. Starting at the bottom, the D class covers insecure systems in which no trust can be placed. An example might be a personal computer in an open office, or a multi-user system to which access can be gained without the need to know a password.

Levels C to A defined progressively increasing levels of security. In level C, the protection features are discretionary. This means that the user or program which owns an object (such as a data file) has absolute control over its access rights, and can grant other users rights such as reading, deleting, amending and extending.

Such discretionary controls can result in a very insecure system if each user chooses to specify lax access permissions on the objects that they own. Level B overcomes this problem by restricting the freedom of users to grant permissions. Mandatory access controls introduce the concept of user security levels. These prohibit a user with a high security level from granting any permissions (at all) to users who have a lower security level.

Such access controls allow the implementation of the 'need to know' schemes required in sensitive applications. These assign to each object and to each user the minimum level of trust needed to accomplish a task. For this reason, it is level B which is of interest to military and government purchasers. The mandatory protection of level B begins to become intrusive. For example, should electronic mail from more trusted users to less trusted users be permitted, and, if so, how should its content be monitored?

The top level, A, does not differ in its facilities from the most secure level B environment. Instead, it carries the concept of verifiability, an increasingly stringent requirement as a system seeks to attain higher levels of trust. In order to satisfy the requirements of level A, a system must not only be secure but its security must be capable of formal proof through an analysis of its software design. It is very difficult to achieve this type of assurance unless the operating system software has been designed with this goal in mind.

Figure A4.2 shows the seven categories currently defined by the Orange Book. The C class is divided into two categories, with class C2 offering more security than C1. Similarly, class B3 allows higher security than B2, which,

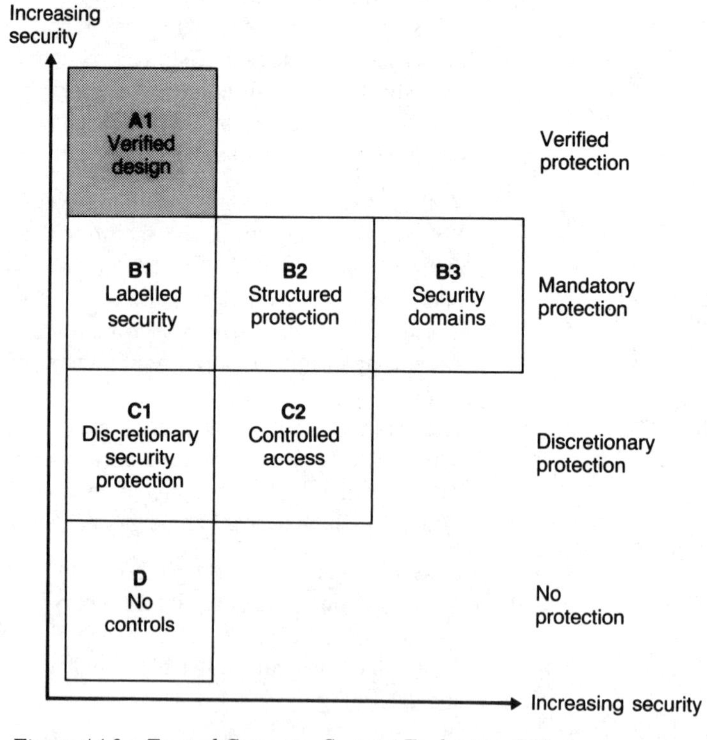

Figure A4.2 Trusted Computer Systems Evaluation Criteria

in turn is more secure than B1. There is currently only one class at level A. Further classes may be added in the future, as more powerful formal methods of verification for software—and for the hardware which runs it—are developed.

The US Department of Defense considers compliance to level B2 or above necessary in systems used in sensitive applications. Such systems are only just becoming commercially available, and less secure products have been pressed into use as an interim measure. Level C2 or B1 should be sufficient for the majority of commercial applications, however.

Briefly examining the salient features of the various levels:

- **Discretionary security protection** encompasses such things as a password-protected log-in mechanism and user-controlled file access permissions.
- **Controlled access permission** brings in more stringent requirements for auditing in order that individual users' actions can be traced—if the system administrator wishes.
- **Labelled security** requires that a *label* specifying level of trust is attached to each security-critical item in a system, in order that the system can evaluate the legality of each operation requested.
- **Structured protection** begins to bring verifiability and documentation to the fore by requiring that the system software is written as largely independent modules, so giving a greater level of assurance that security will not be compromised by unintended cross-module effects.
- **Security domains** amplify the structured protection concept by requiring not just that the system is well-structured, but that the structure is designed with the intention of providing a firm basis for trust.
- **Verified design** requires formal proof of trustworthiness.

The UNIX operating system and the Orange Book

The Orange Book is by no means specific to the UNIX operating system, although enthusiasm for open systems technology (and consequently for UNIX) in the US Department of Defense has resulted in intense development effort being applied to a secure UNIX. There are precedents for this:

- Multics, an operating system dating from the late 1960s, and the main precursor to UNIX, contributed much to the thinking underlying the Orange Book and satisfies the requirements of the B2 level.
- The Honeywell (now Bull) SCOMP, a secure communications processor which currently provides the only environment satisfying class A1 criteria, offers a user interface which derives from that of UNIX, even though the internal design is completely different.

Suppliers' developments have been accelerated because, from the early 1990s, all purchases of computer systems by the US Department of Defense will require satisfactory evaluation at level C2 or above. Other Federal Information Processing Standards (FIPS) will require conformance to the POSIX standards. With hundreds, or even thousands, of millions of dollars in Federal procurements at stake, implementors are rushing to modify their versions of the UNIX operating system, and through the necessary test procedures.

Neither modification nor testing comes cheap however. Informal figures from NCSC quote 2–20 person-years to bring 'vanilla' UNIX System V to the Orange Book's level C, 20–50 for level B, and over 50 for level A.

Confirming these estimates, IBM quotes 19 person-years of effort in a three-year project for the testing of a version of UNIX evaluated to the B2 level. This is not surprising as even a routine test to the comparatively undemanding C2 level takes at least six months.

Figure A4.3 shows the situation. The UNIX operating system, through its user log-in procedures and file access permission scheme, needs little work before it can pass evaluation at the C1 level. However, a poorly administered system can degenerate to the untrustworthy D level. The possibility of security features being misunderstood, abused or ignored in this way underlines the importance of the Orange Book's insistence on adequate documentation and administration procedures. Software features alone cannot provide security. Users and administrators must understand them, and must use them correctly.

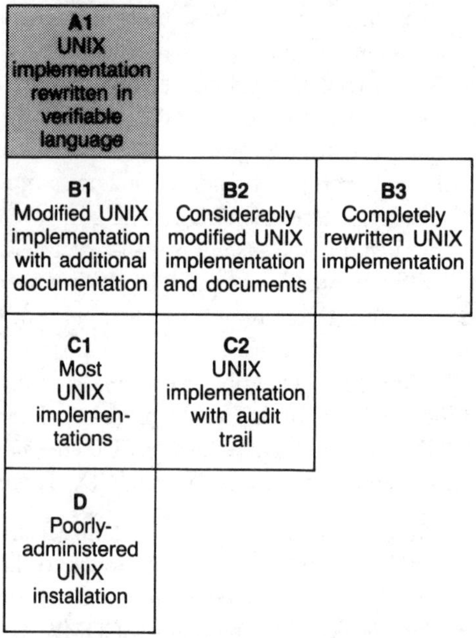

A1 UNIX implementation rewritten in verifiable language

B1 Modified UNIX implementation with additional documentation

B2 Considerably modified UNIX implementation and documents

B3 Completely rewritten UNIX implementation

C1 Most UNIX implementations

C2 UNIX implementation with audit trail

D Poorly-administered UNIX installation

Figure A4.3 UNIX and the Orange Book

Operating systems for self-contained personal computers—MS-DOS, OS/2 and the Macintosh operating system—are unlikely to meet the C1 level. This is because they generally lack a user log-in procedure, and provide file access controls only as an option. File servers might qualify at C1 or C2 level. The OS/2 LAN Manager, for example, implements IBM's mainframe-derived Remote Access Control Facility to provide access control and logging. However, the overall trustworthiness of a system made up of a protected file server and unprotected PCs must be suspect.

Most proprietary operating systems for more powerful computers have, in the terms of the Orange Book, a similar level of security to UNIX. In other words, while UNIX is comparable to other environments in terms of security features, most other environments are, like UNIX, relatively insecure.

It will take a considerable amount of work on AT&T's source code, and on documentation, before UNIX can attain the C2 or B1 levels. Beyond these levels, the difficulty of modifying the original system grows still further. The gap between the B1 and the B2 levels has been described as the biggest in the Orange Book, and results mainly from a required increase in the level of verifiability.

It is feasible that level B2 may be attained by amending the unmodular, twenty year-old code that makes up the UNIX operating system kernel, but the cost may be as much as a complete rewrite. Rewriting is essential if the B3 or A1 criteria are to be met. Both AT&T and the Open Software Foundation are engaged in re-implementing UNIX using modern programming techniques. The results of these long-term projects will be operating systems which can more easily attain the higher levels of trust set out in the Orange Book.

From the point of view of users outside the US, a question mark hangs above the availability of the more secure versions of the UNIX operating system. In the past, the US State Department has required that crypt, a UNIX library routine which implements the NIST Data Encryption Standard (DES), is not shipped outside the US since it is judged to represent strategically important technology. *Crypt* is required in order that user-entered passwords can be checked against encrypted versions held by the computer. UNIX System V avoids the issue by shipping an 'international' version of the routine which encrypts according to the DES, but will not decrypt. The X/Open Portability Guide specifies *crypt*, but is careful not to mention the DES, and carries the warning that 'the values returned . . . may not be portable among X/Open-compliant systems'. In other words, while some implementations may use the DES, others may not.

It is conceivable that the US may place restrictions on the export of secure systems. However, such restrictions are likely only to apply to levels B3 and A1, when conforming systems become available. Systems supporting lower levels of trust—C1, C2, and B1—are already available outside the US without restrictions.

Trusting networked systems

An important aspect of the scheme shown in Figure A4.1 is that, as originally conceived, it applies only to isolated computer systems. The Orange Book makes it clear in its opening pages that networked systems are outside its terms of reference.

The NCSC had not addressed the considerable threats to security presented by remote access from users outside the physically-secured environment of the computer. In other words, a computer system and its operating software (the trusted computer base, in the terms of the book) could be trusted if it was protected not only by its own software, but also by some form of physical access control policy which prevented unauthorized users from communicating with the computer—and so attempting to access it under its layers of software security. Figure A4.4 shows this situation.

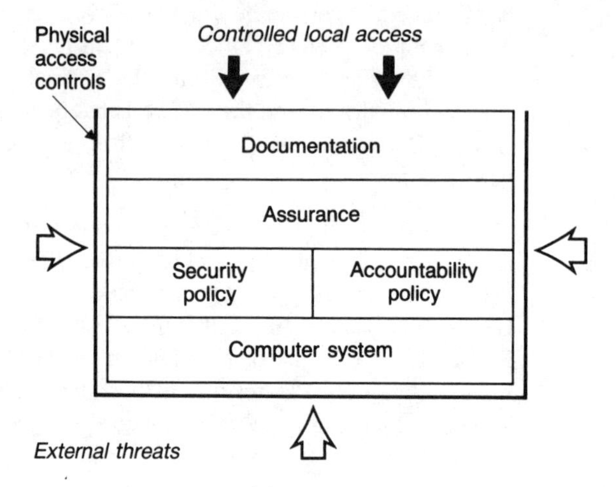

Figure A4.4 Original Orange Book proscribes external threats

As a measure to satisfy the requirements of security-critical military applications, the Orange Book was more than adequate. There was a sufficient market for isolated trusted computers to encourage the implementation of systems which could pass its evaluation criteria, with the result that the requirements of such systems became widely understood.

The issue of systems with external connections remained a problem, however. Was it impossible for such systems to form a trusted computer base? Could the Orange Book be amended to cope with their needs, or would it need complete replacement? Without answers to these questions, the security requirements of most commercial—and many governmental—applications of computers remained unaddressed by any standard. These applications cannot be satisfied by a computer which is isolated in a secure facility. External access is essential for everything from military command, control and communications systems to banking networks.

The direction taken by computer technology since the Orange Book appeared has exacerbated the problem. Personal and workstation computers are increasingly connected to networks as a matter of course and require the services of other computers to get their job done. Without file servers, computer servers, communications servers, and so on, the potential of the computer on the desktop is limited.

While several implementors have come up with trusted operating systems for both desktop and server computers, that trust evaporates as soon as those systems are connected to a network. Developers and users of networked computer technology are becoming increasingly concerned about security issues, and have sought a means of applying the criteria for trusted systems to them, even though they fall outside the terms of reference of the original Orange Book.

In 1989, the NCSC produced the Trusted Network Interpretation (TNI). This says, in effect, that the original Orange Book needs no modification in order to apply to networked systems. All the criteria that it specifies can be applied, and, if they are met, the level of trust which can be accorded to a networked system should be just the same as that for an isolated system which meets the same requirements. (See Figure A4.5.)

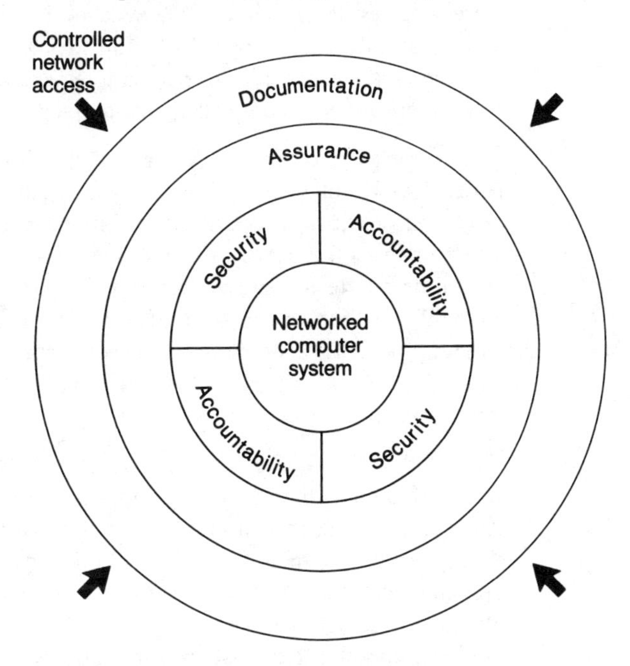

Figure A4.5 Trusted Network Interpretation of Orange Book

This interpretation is both timely and welcome, since it provides a much-needed path to trusted networked systems. But it may be a long path. Issues which can easily be controlled on an isolated system, such as administration, become much more difficult in networked configurations. In this and other cases, new technology must be developed before networked systems can hope to meet the requirements of the Orange Book.

Other issues must be considered on a network. How secure, for example, is data crossing the network? The answer to this question in most cases today is that the data is not secured against unauthorized examination. This and other issues must be addressed before the system as a whole can be trusted.

The interpretation given by NCSC does not mean that a system which has proved that it can be trusted in isolation can be now be connected to a network and retain any level of trust whatever. The whole configuration including both computer and network must be re-evaluated. Because of the technical challenges that this presents, is likely to be several years before any networked system can achieve a high level of trust.

Status of standards for security

It is important to realize that the Orange Book provides a framework for building systems which can be trusted. It does not deal with the implementation of the systems. Those details are left to systems designers. In this, the framework is similar to the Open System Interconnect (OSI) seven-layer model for communications.

The Orange Book model is also similar to the seven-layer model in that increasing levels of trust are achieved by building on top of lower levels—nothing is discarded. However, while the OSI model has existed for some time, and many standards covering implementation details have been derived, as yet there are no security standards derived from the Orange Book.

This situation is changing. As part of the POSIX family of standards, the IEEE POSIX 1003.6 working group is creating a specification for security extensions to the basic operating system interface specified in POSIX 1003.1. The group was formed in 1987, building on work in the Technical Committee of Uniforum. Results are expected from POSIX 1003.6 at the end of 1990, and will take the form of programmatic interfaces to services specified by the Orange Book. IBM, which has invested heavily in secure implementations of its AIX derivative of the UNIX operating system, has contributed much to this work.

The POSIX 1003.6 group liaises with the National Computer Security Commission's own working group on trusted UNIX, known as Trusix. X/Open also has a working group on trusted UNIX. But the current edition of the X/Open Portability Guide does not describe any operating system enhancements in the area of security. Despite this, many of X/Open's members market variants of the UNIX operating system with such modifications. In the absence of standards, there is little compatibility between alternative implementations. The situation is made worse by an absence of standards in the area of system administration, as many aspects of system security concern administrative issues.

Although the NCSC is heavily involved in the specification and evaluation of trusted systems, organizations such as ISO are also working in the field. Emerging OSI standards for network security, and for network management are likely to have an important bearing on the future development of secure systems.

At a different level, many governments support bureaux concerned with computer security in either military or commercial applications. For example, in the UK, the Communications and Electronics Security Group (CESG) and Commercial Computer Security Centre (CCSC) fill these roles.

Summary: security

- It is now possible to purchase systems which pass evaluation at the lower levels of the Orange Book's criteria for trusted systems.

- There are several C2 systems on the market, and B1 systems are just appearing.
- Because testing is such a long process, products may be marketed before formal testing is complete. This means that users requiring a certified level of trust from an environment have less choice than those prepared to take a chance. As an example, the Santa Cruz Operation's implementation of UNIX System V Release 3.2, though not yet certified, is now shipped with C2 security features.
- At higher levels, because of the immense development, documentation, and testing efforts required, commercial products have yet to appear. B2 should be achieved in the early 1990s, but B3 is considerably further out. A1 may just be achieved this century.
- As required security levels increase, more and more features which go beyond those standardized in the POSIX 1003.1 standard are required.
- Today, each secure system implementation uses different techniques to provide these features, with the result that there is little compatibility between systems. Audit trails, labels for resources, and access control lists are examples. All of these pertain to system administration, which is itself a key component of any secure system.
- The whole area of open systems administration has yet to be standardized in any significant way. Thus, administration techniques can vary considerably from one manufacturer's secure system to another's.

The overall conclusion is that it is possible to take a commercially-available open system and modify it to a level of security as good as to the best that can be achieved by any type of system today. The result will be a closed system however, and it will require the use of programming and administrative techniques which are exclusive to that system. This situation will persist until the POSIX working groups come up with the 1003.6 security extensions standard, and the 1003.7 administration standard—and until suppliers start shipping conforming systems. This is unlikely to happen before 1992.

The standards which ultimately appear must be able to accommodate networked systems. As these have only recently been brought within the scope of the Orange Book, and are known to present difficult security problems, any current claims about their level of security should be regarded with suspicion.

In the end, the technology for security can only be an 'enabler'. Effective security will always rely on well-thought-out, well-administrated and well-monitored procedures.

Appendix 5 System administration

Computers require administration. Without it, they cannot give users the service that they require, and they cannot continue to provide service over extended periods of time. This has long been recognized for mainframe computers. Such sophisticated and expensive hardware has always required a number of systems programmers and operators to keep it running, and to adapt it to the changing needs of its users.

Over recent years, techniques for mainframe administration have improved considerably. Increasingly reliable hardware and automated procedures reduce the need for staff to be in continuous attendance, so cutting labour and overhead costs.

Increasing automation in mainframe administration does not mean that administration personnel become less skilled. Newer technologies have emerged that require their attention. For example, network management has come to be a highly skilled (and highly paid) profession. The coordinated administration of open systems, on the other hand, has yet to be recognized as vital, and management tools have yet to be standardized to any significant extent.

The user view

As with so many aspects of the open-systems movement, pressure for standardization of tools for systems administration is coming from large user organizations in general, and from government-backed organizations in particular. It is these organizations which are giving the strongest early backing to one of the fundamental concepts of open systems—that it should be possible to build up networks of systems from multiple vendors, using the same applications software on many of them, and be able to move users from one system to another, without necessarily requiring any retraining.

Superficial examination of the UNIX operating system suggests that system administrators should be 'portable' in this sense, for the following reasons.

- The basic mechanisms for registering, controlling, and logging the actions of users are common to all implementations of UNIX.
- The operating system is always shipped with utilities which can archive and restore data.
- There is an emerging set of common tools for the control of networked systems, due to widespread adoption of Sun Microsystem's Network File System (NFS) and its configuration database.

However, closer analysis identifies two problems.

1. As with most aspects of UNIX, the common facilities offer only a low-level interface.
2. There are many differences between manufacturers' implementations of the, supposedly common, software tools.

One problem can be illustrated, for example, by the long-standing dispute over the relative merits of two file archive mechanisms in UNIX, *tar* and *cpio*. In order to provide a mechanism for source code transfer between systems, the POSIX standard and the X/Open Portability Guide (XPG) define file formats for *both tar* and *cpio*, rather than choosing one or the other. However, in neither case do the standards cater for archives which extend over a number of 'volumes' (diskettes, tapes, etc.). This severely limits the usefulness of the specifications as a basis for high-volume security file back-up procedures.

Print control is an additional problem area. The facilities of the UNIX operating system are very basic in comparison with those available on mainframe systems. For example, there is no knowledge of different types of form, no ability to restart failed jobs in the middle, and no concept of job priority. Provision of such features requires custom coding by a systems programmer, or their simulation by particular applications packages. This limits portability of the software and compounds support problems.

Limitations of this type make the administration of UNIX systems through the use of the basic tools provided with the system very difficult. Although a number of recent books have addressed this issue, the process still requires programming at the level of the UNIX shell command interpreter. Many users are not prepared to do this.

Serious suppliers of UNIX-based computer systems have developed proprietary procedures for the administration of their own hardware. For example, IBM provides common administrative tools for its own range of systems using the AIX operating system, IBM's proprietary version of UNIX. Similarly, a different set of common tools is provided by AT&T in UNIX System V for its own hardware range.

While a user organization with a small number of systems and a single source of supply may be willing to use that supplier's administration procedures, large user sites with necessarily mixed-vendor environments is unlikely to be able to use one particular vendor's administrative tools. Clearly there is a need for standardization of the tools needed.

The problems of administration

Unfortunately, there is a lack of standards in the area of systems administration. Figure A5.1 shows a typical corporate system, consisting of a variety of types of computer and a mix of local- and wide-area communications equipment. Almost everything in the picture needs routine administration of some kind.

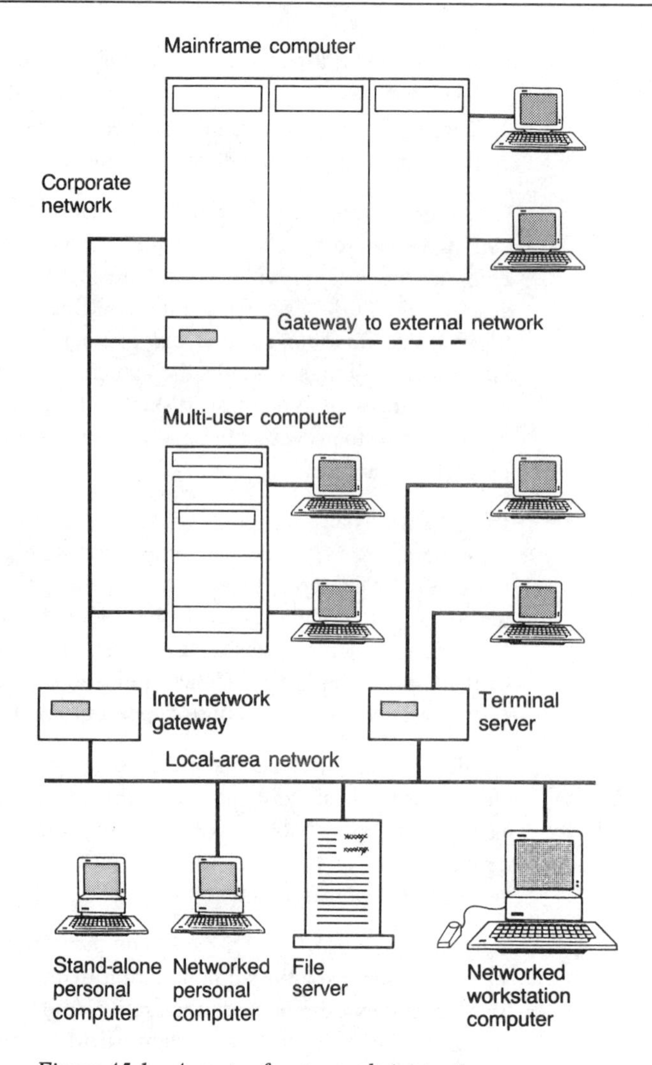

Figure A5.1 Aspects of systems administration

Mainframe computers

Mainframe computer administration is relatively well-defined and mature which, in an increasing number of cases, can be automated to the point of requiring little or no human intervention. In part, this change has been brought about by willingness of users to invest in expensive, but labour-saving, equipment such as mass-storage devices which can replace manually loaded tapes. Another factor is the availability of highly sophisticated software for the remote diagnosis and handling of exceptional situations.

To some extent, this systems management technology has now moved down from the mainframe world on to larger mini-computers. Facilities management companies, specialists in the use of such techniques, have made large and profitable business from running many types of data processing installation under contract for their customers. But the techniques have yet to reach the new world of open systems to any significant degree.

Multi-user computers

Conventional mini- and super-mini computers such as the IBM AS/400 and larger members of the Digital Equipment VAX range running proprietary multi-user operating systems are expected to need the attentions of an administrator.

The UNIX operating system, is a multi-user mini-computer operating system, with corresponding administrative requirements. In its early days as an operating system, UNIX systems were looked after at each site by an expert—a highly qualified person who, by reference to the manuals, could for example devise a made-to-measure back-up strategy, and who was likely to be able to fix any unexpected problem.

With UNIX adopted as a component of an open-systems strategy, a full-time and expensive expert is not an option for most installations. Less qualified staff must bear the load. To achieve this, simplified administration procedures are required.

The authors of applications software which runs on the UNIX operating system usually consider, quite justifiably, that administration is not their responsibility. Few packages therefore embody significant administrative features, on the assumption that a system manager and system tools are available to handle the job.

In fact, lack of standardization at the system level makes it difficult to write portable administration procedures. It is often left to those who sell and install systems to produce customized procedures for their customers.

A small number of independent developers market system administration packages which provide a unified administrative interface to systems from a variety of suppliers. This allows a single set of procedures to be used in a mixed-vendor environment. Take-up of these products has been surprisingly low, however. Either because users are unwilling to pay extra for a service that they think should be an integral part of the basic system, or system vendors do not tell their customers about alternatives to the proprietary administration features of their own offerings.

File and database servers

A network server may be a dedicated system, such as those available from 3COM and Ungermann-Bass, or alternatively, it may be a general-purpose multi-user computer which, among other functions, holds data on behalf of remote systems. In either case, back-up of the data is required, and the identity and privileges of users must be policed. Dedicated servers generally provide well-defined, but supplier-specific, proprietary procedures to handle these functions. Training for administrators is usually offered by server suppliers, but training for one supplier's system may be of little help in the management of another's. General-purpose computers, used as servers, provide facilities which are as good—or as bad—as those described in the previous subsection.

Isolated personal computers

Individual PCs are seldom properly administered. Since the computer is personal, it falls to the person who uses it to perform administrative functions such as file back-up. Few PC users have the discipline to do this job, and many are ignorant of the need for it. The slow speed and low capacity of the floppy disk drives fitted as standard equipment to most PCs serves as a further disincentive.

Some comfort is given by the suppliers of specialized software packages such as accounting systems. Recognizing that the system's back-up facilities are likely to be ineffective and under-used, they provide their own. This protects the data controlled by the package. General purpose software, such as word-processors and spreadsheets, usually lack such features.

While it is possible to configure many PC packages so as to speed particular common operations, most users lack the necessary knowledge to do this, and so waste time by continuing to use the software in an unconfigured form.

There is an argument that if a PC system is badly configured, or if data is lost, only one person suffers, so the damage is more well-defined and limited than for shared resources. However, since damage to the productivity of any individual in an organization is likely to have an effect on the productivity of the organization as a whole, this is not a good argument.

Networked personal and workstation computers

These computers sit between isolated PCs and file servers. If, like many workstations, a particular computer has no disk storage of its own, then it is clear that the responsibility for file back-up rests with the administrator of the server (or servers) used by the computer. The workstation's user may be responsible for its configuration and for aspects of the security of the data that it controls.

Where a computer uses a local disk, as well as remote file storage facilities, it is less than clear where responsibility for back-up lies. It may be possible for the server administrator to use expensive, high-speed, high-capacity back-up hardware to save and restore the contents of the disks on individual users' computers. Even if this facility is available, the security risk of 'publishing' private disks may be judged too great to allow its use. In most cases, it falls to each computer's user to maintain the contents of its private disk, giving rise to the problems discussed in the previous subsection.

Local-area networks

Local-area networks (LANs) lend themselves to management by exception. After the hardware has been set up, and its control software configured, little attention is likely to be required. The issue of administration arises only when the network is extended, when a significant change is made in its usage or when something goes wrong. As with UNIX systems and personal computers, no standard cross-industry procedures exist for administration. Knowledge of particular technologies and particular equipment is essential.

The needs of a small network can probably be handled through a maintenance contract and a good relationship with a supplier or independent network specialist. Larger networks, particularly those involving gateways between network segments, and a variety of technologies (Ethernet and Token Ring, for example) present exceptional circumstances more often, and may begin to merit specialized administration staff.

Wide-area networks

Like the administration of mainframe computers, the administration of wide-area networks (WANs) is big business. Similarly the control and recharging of costs plays an important part in that administration. Commercial products are available from major suppliers, such as IBM and AT&T, together with many offerings from specialists, to serve this market. All are proprietary in nature.

Standards for the management of open (OSI) networks are beginning to emerge. These procedures and protocols will cover the needs of both wide- and local-area networks. Although practical implementations will appear in the near future, it is likely to be several years before these standards are incorporated into commercial products.

Emerging issues

The point of administration

The new methodology of computing brought about by networking raises new issues in administration. With previous generations of hardware, a single central computer communicated with its users through unintelligent terminals (case A in Figure A5.2). In this situation, it is clear that the central computer is the point of administrative control. It is the only configurable element of the system, and allows a single administrator absolute control over the facilities which are made available to the system's users.

A: Central multi-user computer

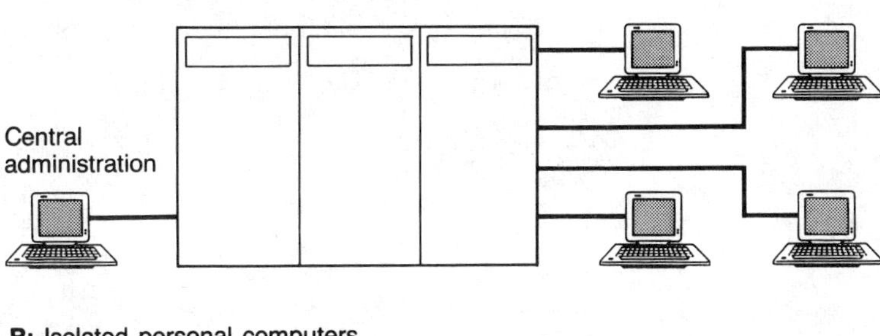

B: Isolated personal computers

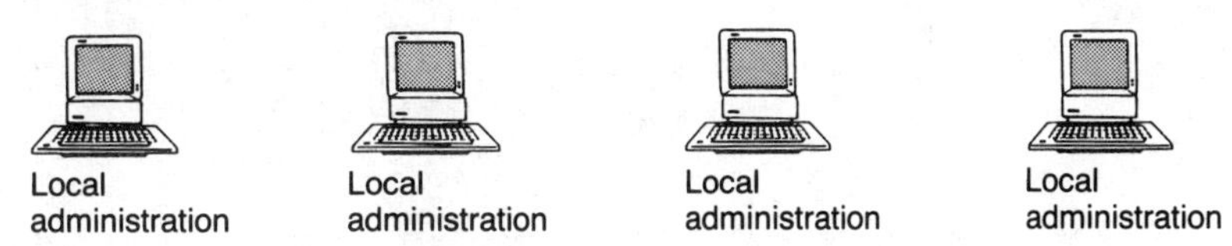

C: Distributed processing environment

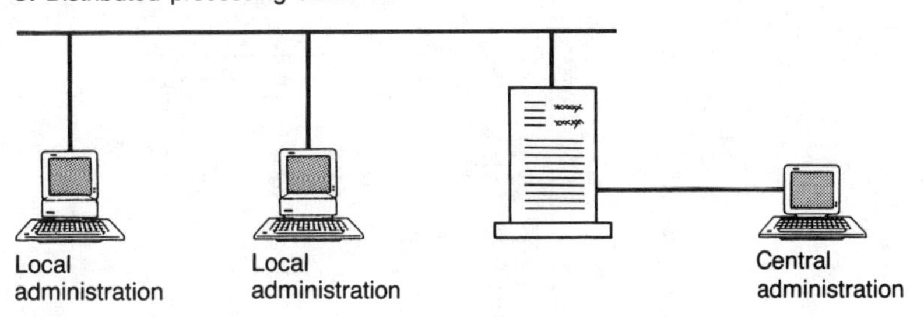

Figure A5.2 Location of administrative control points

Isolated personal and workstation computers present a similar picture (case B). There is a single point of control on each computer, and it determines the facilities offered to the user of that computer (who is generally also responsible for its administration).

Only when computers are linked over a network does the situation change substantially (case C). In addition to the intelligent devices in front of each user, there are also computers acting as shared resources. Each element of the distributed processing environment requires administration, and the administrative tasks fall into two parts: those which affect only one user, and those which affect more than one.

It is possible, in a networked environment, to concentrate all administration to a single point, creating a situation like that in case A and denying users any control over their own environment. It is usually judged better, however, to allow users to customize the configuration of their own systems, perhaps so as to increase their own productivity, provided that the overall integrity of the distributed environment is not compromised. Thus, each user is responsible for elements of local administration, while a network administrator looks after shared resources.

Networks have a tendency to become larger in extent, and in their populations of users. It is convenient, both technically and administratively, to split a large network, or *domain*, into *sub-domains*, as shown in Figure A5.3. This division adds another tier to the control structure:

- Users administer their own systems

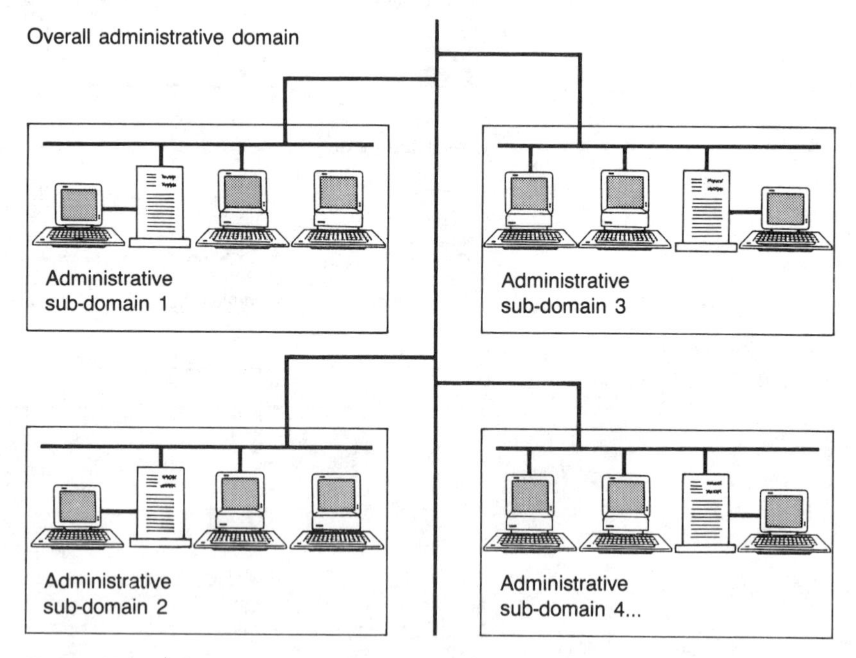

Figure A5.3 Administration in a large distributed environment

- Local administrators manage each domain
- An overall control function manages global aspects of the network

The process can be repeated several times, with the top-level domain in the figure becoming a sub-domain of some yet larger network, and so on.

Any proposed standard for system administration must accommodate this model of the world, and yet be comprehensible and manageable at each level. This is one of the main reasons why standards will take a long time to appear.

Distributed software

Software package installation and policing is another issue which grows in importance and complexity in networked environments. 'Software piracy', i.e. illegal copying and use of software in contravention of the terms of the licence agreement between the vendor and purchaser, is a serious concern. Networked systems make it extremely easy to copy software from one system to another, or to use a program which resides on a remote disk. Indeed, a workstation computer without its own disk is obliged to use one of these mechanisms in order to run any program at all.

Software authors are seeking some means of limiting the usage of their packages once they have been installed on a system which is part of a network. System administration standards are expected to address this issue, together with that of ensuring that each package is completely and correctly installed in the first place.

Recent security threats of tampering, worms and viruses have heightened concern for the integrity of software packages once installed. Again, a comprehensive administrative standard should take steps to protect installed software, and to detect possible cases of unauthorized access.

Standards work in progress

The situation today regarding standards for the administration of open systems is that a number of products exist in the commercial arena, and a great deal of research is under way, but no clear standards are emerging. Formal standardization efforts have only just started with a view to producing results by late 1991.

Commercial implementations

Together with other manufacturers, both AT&T and IBM provide proprietary administration tools for their UNIX implementations. The status of both sets of tools is changing in ways which could result in either—or both—reaching *de facto* standard status.

Prior to the release of UNIX System V Release 4, AT&T included only the low-level tools mentioned earlier. The higher-level administrative tools were not included, and were shipped only with binary releases developed by AT&T for its own computer hardware products. This meant that other vendors, in porting the UNIX operating system to their own hardware, had to develop high-level tools of their own. Consequently, no two suppliers' systems offered the same high-level administrative interface, although knowledgeable users could find common low-level tools if they wished.

For UNIX System V Release 4, AT&T took the decision to ship an improved version of its Administrative User Interface with the source code (Figure A5.4). In order to encourage the adoption of its system, AT&T has made it extensible, so that services not catered for in the basic release can be added. Whether its customers will use this is open to question, as most by now have highly developed proprietary administrative interfaces, and may be reluctant to replace them, or to support AT&T's solution in addition to their own.

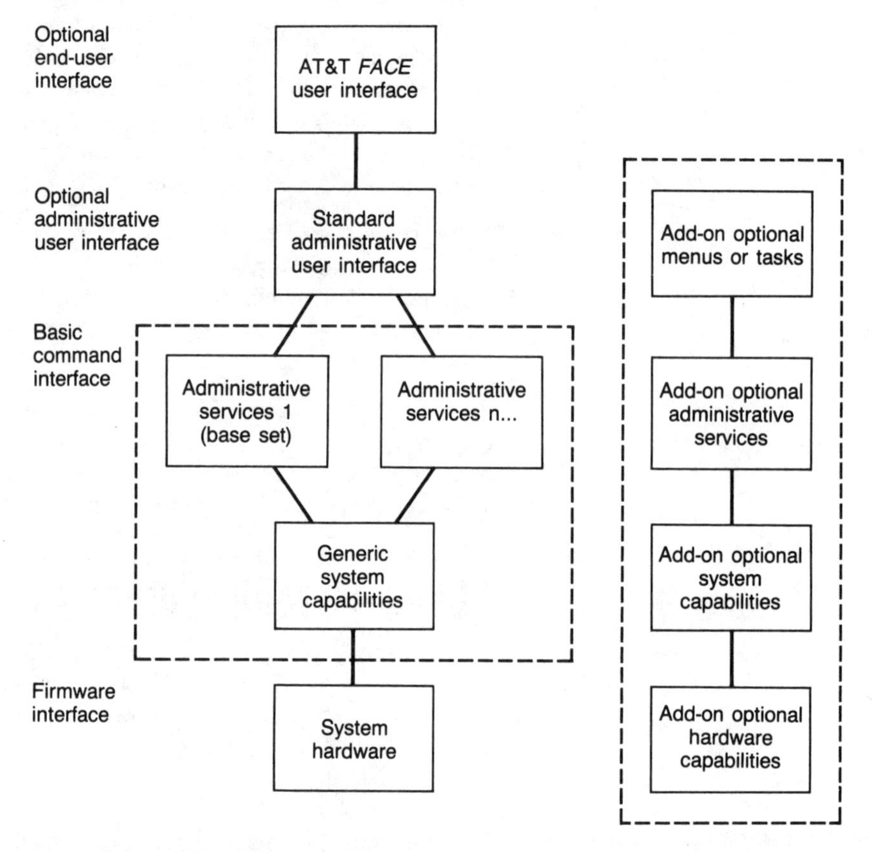

Figure A5.4 UNIX System V Release 4 administrative services

IBM's administrative system for its AIX operating system assumes a networked environment, unlike AT&T's tools, which add network administration features to a single-system management scheme. The first release of the Open Software Foundation's operating system, OSF/1, due for release in mid 1990, may include the AIX administrative tools. If this is so, they will join AT&T's offerings in the race to become a *de facto* standard.

Academic research

Two major US academic projects involve system administration as part of their brief. Both Project Athena at the MIT and the Andrew Project at CMU are investigating the problems of managing large networks of interconnected workstation and server computers. To give an idea of the scale of the projects, both are concerned with ultimate computer populations of well over 1000, and user populations exceeding 10 000. The projects are heavily funded by industry; IBM, for example, is involved in both.

Project Athena, is exclusively based on the UNIX operating system running on a variety of types of computer. The Andrew Project has a wider brief, accommodating Apple Macintoshes, IBM PC-compatibles, and several other types of system, in addition to those running UNIX. It aims to provide a unified environment across them all.

A primary aim in both projects is to ensure that the administrative load presented by the network grows more slowly than the number of systems connected. Since both cater for the needs of students, the task of administration is difficult. Users must be allowed full control of their own computer, yet any damage that they do as a result of this freedom must be localized and amenable to speedy repair by a remote operator or an automatic process. For example, if the contents of a workstation's disk are destroyed, the contents must be restored over the network. Similarly, security is of vital importance. Sensitive data must be protected from those people who are not authorized to use it, but who may make determined attacks (as students are likely to do) on the system in order to obtain it.

Apart from the X Window System from MIT, little from either project has yet appeared in commercial implementations of the UNIX operating system. Both are likely to contribute significantly to many areas of open system technology, including that of system administration. In particular, the *Hesiod* name server (akin to an electronic telephone directory service) and the *Kerberos* authentication service (a comprehensive security mechanism) from MIT are beginning to see widespread use in the US academic community, and may well influence the standards which eventually emerge.

A feature common to the Athena and Andrew Projects is the use of the TCP/IP family of communications protocols. OSI protocols were not available at the inception of the projects in the early 1980s and TCP/IP was the protocol of choice in the US academic and scientific community. As the close of this chapter will show, formal standardization efforts for system administration procedures are beginning to draw on emerging standards for OSI administration, and it might be thought that this would limit the usefulness of experience gained using TCP/IP.

Developments at the National Science Foundation, a US body which funds long-term research, show that this need not be the case. Simple Network Management Protocol (SNMP), a recent addition to the TCP/IP arsenal, is written in terms of ASN.1, a special-purpose language created to describe OSI protocols. Thus there is a point of contact between the two communications methods.

Standards status for systems administration

In early 1989, the US National Institute of Standards and Technology, NIST, the body responsible for the development of US Federal Information Processing Standards (FIPS), circulated a draft FIPS covering the area of system administration. The document was based on existing practice derived from UNIX System V, Berkeley UNIX, IBM's AIX and a number of other implementations. Sensing a serious deficiency in the area of standards for administration, NIST's aim was to produce an approved FIPS by early 1990. This would be binding on suppliers wanting to tender for US government business, and so would be a very strong force for standardization.

The draft FIPS was reviewed by members of the POSIX 1003.7 working group on system administration (see next subsection). Discussing its contents with members of NIST's staff, they reached a consensus that the draft could not form the basis of a useful standard, as many aspects of the existing practice on which it was based were seriously deficient. The control of networked systems and of security gave rise to particular concern. As a result of the discussion, the draft FIPS was withdrawn, pending the outcome of early work by the POSIX 1003.7 group. NIST has tacitly accepted that 1990 is too early a target for a usable standard, but is actively encouraging the IEEE Working Group in its quest for results as soon as possible.

POSIX

Working group 1003.7 of the IEEE was formed in mid 1988 to begin work on a POSIX standard for system administration. Among possible areas for standardization were system start-up, process management, network management, software licence management, user management, and many others—far more than could be addressed in the proposed timescale of two to three years.

Existing practice in the administration of the UNIX operating system was judged unhelpful as a basis for a lasting standard. Instead work is proceeding based on emerging OSI standards for open networks. This means that system administration in effect becomes an extension of network management, and uses common procedures where appropriate.

Among the ISO standards and draft proposed standards being used by POSIX 1003.7 are: IS 8824:1987 (Specification of Abstract Syntax Notation 1—ASN.1), and DPs 9596-1 and -2 (Common Management Information Service—CMIS, and Common Management Information Protocol—CMIP, respectively). The use of ASN.1 is particularly significant, since it is the formal specification language of OSI, and underlines the increasing commonality between the POSIX and OSI standardization efforts.

Batch processing is not considered to be part of POSIX 1003.7's work. Instead, it falls to another POSIX working group, 1003.10, Supercomputing Application Environment Profile, to standardize this aspect of POSIX-compliant environments. Supercomputers running variants of the UNIX operating system already have facilities to cope with batch jobs, so it seems appropriate that the group should create a standard as part of its work in defining the POSIX facilities required for supercomputing. There is no timetable for the production of this standard, however.

Existing standards for systems administration

There are no existing standards for system administration. Existing standards covering other aspects of open operating systems are very careful to avoid administrative issues. This avoidance is both correct and conventional. Formal standards are expected to address just one area, and not to stray into territory which should properly be the subject of another standard.

The only approved POSIX standard, 1003.1–1988, defines almost no privileged aspects of the operating system. The operations that it describes are available to all users. Since administration involves privileged operations it is outside the scope of 1003.1, which does not define the low-level system facilities needed to support administrative functions. This task falls to working groups 1003.6 (security) and 1003.7 (system administration).

X/Open's XPG also omits reference to administration, although some privileged low-level system services are defined. Confusingly, *admin*, a user-level command described in the XPG, is concerned with version control for text files, not with system administration—the command name is a historical hangover. Only the System V Interface Definition (SVID) defines a full set of privileged interfaces, and an Administered Systems Extension. The extension defines 57 administration utilities which form an optional part of UNIX System V, and constitutes the only *de facto* standard in the field.

Summary: system administration

- System administration is one of the least standardized and most problematic areas for open-systems implementation.
- Effective administration of all but the simplest installations is likely to require the attentions of a skilled programmer for set-up, and skilled operations staff for running.
- The standards which will change this situation are in preparation, but will take time to arrive. They must cope with the new concept of networked systems, and so can build only to a limited extent on the years of experience with centralized computer installations.
- Until standards do arrive, and are widely implemented, users of open systems must accept that administration issues must be considered anew for each new machine that they install.

Glossary

ASCII American Standards Code for Information Interchange.

ANSI American National Standards Institute, US, ISO member.

ANSI-SQL Structured query language standard defined by American National Standards Institute.

ANSI-C 'C' language standard defined by American National Standards Institute.

ABI Application binary interface—allows binary portability between machine using the same processor.

AUI AT&T's Administration User Interface.

ASN Abstract syntax notation used in defining OSI protocols.

ARPAnet Connectionless network developed in the US.

ANSA Advanced Network Systems Architecture—a UK government sponsored project investigating ways to produce generalized distributed systems.

BIOS Basic input/output system.

Binary portability Directly executable code which can be moved from one machine type to another, i.e. does not require recompiling or porting.

Bridge A device that connects local area networks which have differing data link layer protocols.

Bus A communications pathway within a computer.

Crypt A UNIX library routine for encrypting and decrypting text.

CCTA UK Central Computer and Telecommunications Agency, which provides advice and guidance to UK government departments to assist with procurement of IT systems.

CALS Computer Aided Acquisition and Logistics Support is a program mounted by the Department of Defense in the US to create a partially paperless Pentagon.

CCITT Committee Consultitif International Télégraphique et Téléphonique, this committee produces recommendations for public telecommunications services.

CBEMA Computer Business Equipment Manufacturers' Association is an accredited standards-making body. Its members are producers of computers, business equipment, services and supplies.

Curses A library of programs within UNIX which allows programs to treat the screen as a series of windows containing characters.

Cpio File archive mechanism.

COS Corporation for Open Systems is a membership based non-profit research and development consortium, comprising computer and communications vendors and major users. It is dedicated to accelerating the introduction of OSI and ISDN products and develop conformance testing for networking standards.

CEN Comité Europeén de Normalisation—European standardization body.

CENELEC Comité Europeén de Normalisation Electronique—European standardization body for electrotechnical matters.

CAE Common applications environment defined by X/Open and modelled on the UNIX operating system. It is designed to ensure portability of application programs at the source code level.

CSMA/CD Carrier sense, multiple access collision detect—the method of arbitration on Ethernet networks.

CIM-OSA Computers for Integrated Manufacturing-Open Systems Architecture, one of the projects under the ESPRIT program.

CNMA Computer Networks for Manufacturing Automation, one of the projects under the ESPRIT program.

***De facto* standard** Usually a product that has been widely accepted and therefore becomes a 'standard' but not controlled by an official standards body.

***De jure* standard** A standard that has been officially approved by a recognized standards body.

DoD Department of Defense in the US.

DARPA US Defense Advanced Research Project Agency.

DIN Deutsches Institut für Normung, German standards body—ISO member.

Distributed processing Sharing of processing tasks between computer systems.

Debugging Identifying and removing faults in software.

DES Data encryption standard—a method of encrypting data.

EUUG European Unix Systems User Group—a non-profit organization bringing together 4000 members of national users groups in sixteen countries. It oversees the management of EUNET, the European segment of the UNIX mail network and also provides joint funding for monitoring of ISO POSIX standards project.

EBCDIC Extended binary coded decimal interchange code.

EDI Electronic data interchange is a set of standard data formats for information exchange.

EC European Community.

ETI Extended terminal interface.

ENs Europėenes Normes—European Standards.

Ethernet Local area network using CSMA/CD techniques.

ETSI European Telecommunications Standards Institute established in 1988 to handle standards for European customer equipment used on the Integrated Services Digital Network.

EMS Electronic mail system.

ESPRIT European IT research program.

FMLI Forms and menu language interpreter.

FIPS Federal Information Processing Standard.

FTAM File transfer access and management, the OSI application layer standard which allows files to be transfered and accessed from remote systems.

FDDI Fibre distributed data interface is a standard for communicating over fibre-optic cable.

Gateway A link between two networking architectures which translates messages from one to the other by passing them through the application layer.

GOSIP Government OSI profile—standards defined by both UK, US and other governments to simplify, purchasing of open systems by government departments.

Glyphs Human-readable representations of characters.

Graphics tablet A device for inputing abitrary shapes into a computer.

GUI Graphical user interface.

GEM Digital Research's graphical user interface.

Groupware Software specifically designed for task sharing among groups of people.

IEC International Electrotechnical Commission.

IMS Information Management System from IBM.

ISAM Index sequential access method is a simple data base access technique.

ISDN Integrated Service Digital Network is an international network that allows movement of large amounts of data at any one time. It provides a number of digital services including voice, video, data and fax.

IEEE Institute of Electrical and Electronic Engineers, an accredited body in the US that creates standards for ANSI. They set up and manage a number of working committees, one of which was POSIX.

ISO International Standards Organisation incorporated by the United Nations. ISO authorize international standards. This is done via industry organizations in each country.

Interoperability The ability to communicate between dissimilar systems in such a way that the characteristics of the system providing the service to the user are transparent.

JISC Japanese Industrial Standards Committee—ISO member.

JIT Just-in-time, a technique used by manufacturers where goods are ordered and supplied only as needed.

JTM Job transfer and manipulation is the ISO application layer standard which will allow jobs to be submitted to remote machines over an ISO communication link.

LAN Local area network is a system providing data communications over a short distance, e.g. a building.

'Look and feel' The general appearance and method of use which is characteristic of a particular graphics user interface.

MIT Massachusetts Institute of Technology—formed the X Consortium and developed the X Windows product.

MAP Manufacturing Automation Protocols is an OSI based functional standard set by the MAP Users' Group. It was originally specified by General Motors when they implemented an open systems policy. GM heads up the MAP Users' Group.

MAC Medium access control method, a sublayer in the data link layer which controls access to the physical medium of a network.

MTA Message transfer agent, the part of an electronic mail service which stores and forwards mail to other systems.

MSS Manufacturers' messaging standard, a messaging system which forms part of MAP.

Mouse A device for directing movement of a screen based pointer.

Medium The physical connection on which a network is based.

NIST National Institute of Standards and Technology, an agency of the US government which specifies US government procurement rules. NIST is responsible for the development of Federal information processing standards.

NFS Network file system developed by Sun Microsystems for distributed file access.

NCA Network computing architecture from Apollo (now part of HP) and fast emerging as the *de facto* standard for remote procedure calls.

NCSC National Computer Security Commission, the US

government department which is the driving force behind most of the recent developments on security of computer systems.

NATO North Atlantic Treaty Organisation, particularly interested in interoperability of military systems of member countries.

NeWS Sun Microsystem's Network-extensible Window System.

Node A logical point of connection to a network, which could be a terminal, printer or computer.

NeXT New hardware company formed by Steve Jobs (ex-Apple), and the machine range of the same name.

OSF Open Software Foundation, a non-profit organization whose aim is to provide the computer industry with a common operating system across all computer hardware platforms. OSF members total more than 150 and cover the whole spectrum from users to manufacturers.

ONA Open network architecture.

ODA Office Document Architecture, the standard produced jointly by the CCITT and ISO which defines the basic structure of text in a document which may be sent by OSI systems.

ODIF Office Document Interchange Form.

OSI Open Systems Interconnection, a set of standard communication protocols, defined in the OSI reference model to be independent of any one vendor, which will allow systems to be constructed which are guaranteed to be able to communicate easily.

OSI reference model The generic model by which open systems are constructed in OSI. Seven separate layers handle the communications process, through standard interfaces. This is also known as the ISO 'seven-layer model'.

OSF Motif The Open Software Foundation's graphical user interface.

Open Look Graphical user interface developed by Sun Microsystems and AT&T.

ODP Open distributed processing.

POSIX The name given to the IEEE 1003 committee and the related standards for defining portable operating system interfaces and functions.

People portability The ability to move people easily between different systems and networks without the constant need for retraining.

Proprietary system A product range of hardware or software controlled by the manufacturer. In the case of hardware, uses its own operating system. Therefore, unless products are ported to this operating environment, the users' choice of software is limited and so is their ability to communicate with other machines from different manufacturers.

Porting The function of making software available under various operating system environments.

PSI Portability, scalability and interoperability, the three requirements for open systems.

Portability The ability to move a piece of software from one manufacturer's machine to another's.

Pixel The smallest displayable element on a screen.

Profile In OSI terms, it is a grouping of OSI standards tailored to meet specific application requirements.

PM Presentation Manager, the graphical user interface from Microsoft designed for OS/2.

PAD Packet assembler/disassembler, a device which converts messages into packets and vice versa on a packet-switched network.

PTT A national postal, telegraphy and telecom agency, e.g. British Telecom in the UK.

PARC Xerox' Palo Alto Research Centre.

RISC Reduced instruction set computer, a computer with a small number of optimized instructions with the potential for high performance.

RPG Remote procedure calls, request processes on remote machines as if they were local.

Raster graphics Images defined by dots.

RFC Requests for comments, part of the process prior to ISO authorization of a standard.

RACF Remote access control facility from IBM.

R&D Research and development.

SQL Structured query language.

Scalability The ability to run the same software with acceptable performance on any size of system.

SNA System Network Architecture, IBM's proprietary network protocols.

SAA Systems Application Architecture, IBM's strategic concept for designing and developing applications to run on a wide variety of machines.

SWIFT Society for Worldwide Inter-bank Funds Transfer.

SCOMP A secure communications protocol.

SNMP Simple network management protocol.

SPARC Scalable programmable microprocessor architecture designed by Sun Microsystems with the intention of making it a *de facto* standard, and licensed to other manufacturers.

SVID AT&T's UNIX System V Interface definition.

Tar A file archive mechanism contained within UNIX.

Track-ball Device similar to a mouse.

Termcap Provides a means of access within UNIX to the individual capabilities of terminals.

Terminfo Terminal information library in UNIX System V.

Toolbox Apple's Macintosh high-level user interface toolkit.

TCSEC *Department of Defense Trusted Computer Systems Evaluation Criteria.* Published by the NCSC, this booklet (known as the Orange Book) plays a central role in the specification of secure computer systems.

TNI *Trusted Network Interpretation*, a document produced by NCSC saying in effect that the specifications contained in the 'Orange Book' will apply equally to network security.

T1 Is a 1.54 Mbyte communications link, often referred to as 'kilostream'.

TOP Technical and Office Protocol, the functional standards promoted by Boeing Aircraft Corporation for the office environment. Closely related to and complementary to MAP.

TP Transaction processing.

TCP/IP Transmission Control Protocol/Internet Protocol, a *de facto* standard approximating layers 3 and 4 of OSI and developed by the US Department of Defence.

UNIX AT&T's operating system, widely adopted by the major hardware manufacturers. It has become a *de facto* standard.

User interface The part of the operating environment that the user has direct access to, allowing him/her to perform tasks by selecting an icon that would previously have required lengthy commands. The user interface usually supports graphics and windowing facilities.

UA User agent.

USENIX Is an organization of AT&T Licensees, sublicensees and other professional and technical persons, formed for the purpose of exchanging ideas about UNIX and similar operating systems and the 'C' programming language. USENIX supplies joint funding (with EUUG) for monitoring of the ISO Posix standard projects.

Uniforum Previously known as /usr/group, this group pioneered much of the work that led to the Posix standard. It is a non-profit organization of open-systems professionals and disseminates information on open-systems technologies through a series of activities, the most well-known of which is the annual Uniforum Conference and Trade Show, held in the US. Uniforum has a strategic partnership with X/Open.

X/Open X/Open is an international non-profit organization set up to define, promote and supply open-systems technologies. It is owned by a consortium of the world's largest computer manufacturers.

Sources for further information

Further information on many of the activities described in this book can be obtained from the following organizations:

American National Standards Institute (ANSI)
1430 Broadway
New York 10018
USA

Telephone: + 1 212 354 3300
Fax: + 1 212 308 1286

Commission of the European Community (CEC)
200 Rue de la Loi
B-1049 Brussels
Belgium

Telephone: + 322 235 1270

European Committee for Standardization (CEN)
Rue Brederode 2, Bte 5
B-1000 Brussels
Belgium

Telephone: + 322 519 6811
Fax: + 322 519 6819

European Committee for Electrotechnical Standardization (CENELEC)
Rue Brederode 2, Bte 5
B-1000 Brussels
Belgium

Telephone: + 322 511 7932
Telex: + 26 257 CENELEC B

Corporation for Open Systems International (COS)
Suite 400
1750 Old Meadow Road
McLean, VA 22102
USA

Telephone: + 1 703 883 2700
Fax: + 1 703 848 4572

Europe Computer Manufacturers Association (ECMA)
114 Rue du Rhone
CH 1204 Geneva
Switzerland

Telephone: + 412 235 3634
Fax: + 412 286 5231

EurOSInet
Secretariat, c/o Level 7
Centenial Court
East Hampstead Road
Bracknell RG12 1YQ
UK

Telephone: + 44 344 867199
Fax: + 44 344 878442

European Unix Systems User Group (EUUG)
Owles Hall
Buntingford
Hertfordshire SG9 9PL
UK

Telephone: + 44 763 73039
Fax: + 44 763 73255

European Workshop for Open Systems (EWOS)
Rue Brederode 13, 2nd floor
B-1000 Brussels
Belgium

Telephone: + 322 511 7455
Fax: + 322 511 8723

European X User Group (EXUG)
185 High Street
Mitchell House
Cottenham
Cambs CB4 4RX
UK

Telephone: + 44 954 211860

International Data Exchange Association (IDEA)
68 Avenue d'Auderghem, Bte 34
1040-Brussels
Belgium

Telephone: + 322 736 9715
Fax: + 322 736 9821

Institute of Electrical & Electronics Engineers, Inc. (IEEE)
345 East 47th Street
New York NY 10017
USA

Telephone: + 1 201 562 3809
Fax: + 1 201 981 1686

International Organization for Standardization (ISO)
General Secretariat
Case Postale 56
3 Rue de Varembe
CH-1211 Geneva 20
Switzerland

Telephone: + 412 234 1240
Fax: + 412 233 3430

Japanese Standards Association (JIS)
4-1-24 Akasaka
Minato-ku
Tokyo 107
Japan

Telephone: + 813 583 8001
Fax: + 813 586 2029

Japan UNIX Society (JUS)
Towa-Hanzomon Corporation Building 505
2-12 Jayabusacho
Chiyado-ku
Tokyo 102
Japan

Telephone: + 815 561 3068

Marosi Ltd
Scammell House
High Street
Ascot SL5 7JF
UK

Telephone: + 44 990 873155
Fax: + 44 990 27325

Ministry of International Trade and Industry (MITI)
3-1 Kasumigaseki 1-chome
Chiyoda
Tokyo 100
Japan

Telephone: + 813 501 1657
Fax: + 813 501 2081

National Institute of Standards and Technology (NIST)
Technology Building
Room B266
Gaithesburg MD 20899
USA

Telephone: + 1 301 975 3295
Fax: + 1 301 590 0932

Open Software Foundation (OSF)
11 Cambridge Center
Cambridge MA 02142
USA

Telephone: + 1 617 621 8748
Fax: + 1 617 225 2943

Open Systems Interconnection/Technical and Office Protocol (OSITOP)
CIGREF
21 Avenue de Messine
F-75008
Paris, France

Telephone: + 33 147 642 848
Fax: + 33 147 642 579

Sigma Organization (SIGMA)
Senwa-Toyo Building, 6th Floor
3-16-8, Sotokanda
Chiyoda-ku
Tokyo 101
Japan

Telephone: + 813 255 0421
Fax: + 813 255 0473

Standards Promotion and Applications Group (SPAG)
Avenue Louise 149
Box 7
1050-Brussels
Belgium

Telephone: + 322 535 0811
Fax: + 322 537 2440

UniForum
Suite 201
2901 Tasman Drive
Santa Clara CA 95054
USA

Telephone: + 1 408 986 8840
Fax: + 1 408 986 1645

Unix International, Inc. (UI)
Waterview Corporate Center
20 Waterview Boulevard
Parsippany NJ 04054
USA

Telephone: + 1 201 263 8400
Fax: + 1 201 263 8401

Unix Software Operation
International House
Ealing Broadway
London W5 5DB
UK

Telephone: + 44 815 677711
Fax: + 44 815 672420

USENIX
PO Box 2299
Berkeley CA 94710
USA

Telephone: + 1 415 528 8649

The X/Open Company Ltd.
Apex Plaza
Forbury Road
Reading RG1 1AX
UK

Telephone: + 44 734 508311
Fax: + 44 734 500110

Index

ABI, 90
Abstract Syntax Notation (ASN), 104
Access control standards, 39
Accredited bodies, 52
ADA, 31, 207
Addressable elements, 199
Administration, *See* System administration
Adobe Systems Inc., 48
Advancement, speed of, 5
AFCAC 921, 58
AIX, 239, 246
Alvey Mail, 159
Alvey research initiative, 119
Amadahl Executive Institute, 2
Andrew Project, 247
ANSA, 119
ANSI, 52, 55–58, 62, 65, 68, 69, 107, 124, 175–177, 179, 191, 192, 220
Apple Macintosh, 30, 66, 71–72, 167, 188, 200, 201, 206, 220, 222, 232
Application Binary Interface (ABI), 86–89
Application Binary Interfaces (ABI) standards, 35
Application environment standards, 66–74
Application generators, 32, 198
Applications programming interface (API), 68–69, 147
Architecture Neutral Distribution Format (ANDF), 83, 85
ASCII, 54, 73, 74, 167, 220, 222
Ashton-Tate, 185
Asia-Pacific Telecommunity (APT), 118
Asian ISDN Council (AIC), 118
ASN.1 language, 247
AT & T, 56, 58, 66, 70, 74–76, 80, 82, 90, 195, 208, 233, 245
Audio tape cassette, 47

Backlogs, 1, 2
Balloting groups, 61–63
Balloting procedure, 63
BASIC, 31
Benefits, 3
Beta, 9
Betamax, 48, 49
Binary compatibility, 86–94
Binary portability, 34
BIOS, 49
Bit-maps, 206–207
Boyle, Brian, 79
Branding, 40
Bridges from proprietary to open systems, 37
British Airways, 151–153
BSD UNIX, 50
Business professionals, 139

C-ISAM, 178, 179
C language, 31, 53, 68, 69, 74, 124, 176, 181, 207, 220, 224
C++ language, 68, 207
C programming language, 56, 178
CALS, *See* Computer-Aided Acquisition and Logistics Support
Carnegie-Mellon University (CMU), 72
CCITT, 53, 103, 108, 115
CCTA, 113
CD-ROM, 73
CEN/CENELEC, 130
Character representations, 218
Character sets, 42, 218–221, 224
Character terminals, 198
  look and feel for, 197
  standards, 191–192
Character User Interface Style Guide, 197
China, 42
CIM-OSA, 119
Client programs, 204
Client-server window architecture, 203–205
CNMA, 119
COBOL, 31, 53, 68, 69, 124, 176, 178, 179, 181, 182, 184
COBOL-85, 124
Codd, E. F., 174, 175
Collating sequence, 221, 225
Commercial Computer Security Center (CCSC), 236
Commercial users, incompatibility issues concerning, 16
Common Applications Environment (CAE), 122–126
Common language, 97
Communications and Electronics Security Group (CESG), 236
Communications standards, 51
Compact cassette (CD), 47
Compact disc, 51
Compatibility, 6, 8
  advantages of, 34
  binary, 86–94
  upward, 89–90
Compatible components, 25
Compatible systems, 5
Competitive advantage, 22–23
Competitiveness, 5
Compilation, 34
Computer-Aided Acquisition and Logistics Support (CALS), 51, 117

Computer-aided design, 25
Computer-aided software engineering (CASE), 185
Computer hardware, *See* Hardware
Computer industry, 5
Computer users, incompatibility problems concerning, 14
Conformance testing, 114
Conformance testing and branding, 39–40
Connected islands, 169
Controlled access permission, 231
Copyright law, 30, 49
Correspondents, 62
COS, 6, 66, 144
Costs, 3, 4, 19
Creativity, 16, 34
Crypt, 233
CSMA/CD, 106
Curses, 194, 195
Customer demands, 138
Customer power, 138–139

DARPA, *See* Defense Advanced Research Project Agency
Data distribution, 2
Data Encryption Standard (DES), 233
Data format, 172–173
Data handling, 167–186
  standards, 171
Data management, 69–70
Data standards, 167
Database engine, 174–175
Database management systems, 32, 167–168, 170
Databases, 218
  organization-wide, 170
  system administration, 241
Date and time, 222, 225
DB2, 69, 171, 176
dBASE, 185
De facto standards, *See* Standards
DEC, 67, 73
DEC PDP-11, 75
Defense Advanced Research Project Agency (DARPA), 51
Delay times, 2
Department of Defense (DoD), 51
Department of Trade and Industry (DTI), 3, 158–164
Development costs, 19
DHL, 149–151
Digital Equipment Corporation, 67
DIN, 52
Directory management functions, 111–112
Discretionary security protection, 231
Display PostScript, 71
Distributed database, 180
Distributed processing, 29
Distributed software, 245
Distributed systems, 25
Distribution media, 90
DROP statement, 177

Earl, Michael, 1
EBCDIC, *See* Extended Binary Coded Decimal Interchange Code
EDI, *See* Electronic data interchange
Electronic data interchange (EDI), 36, 66, 96, 111, 115–117
Electronic-mail systems, 110–111, 158–164
Encryption of data, 39
English as standard language, 42
Environment variables, 217
Error messages, 223
ESPRIT research program, 119
Ethernet, 103, 106, 157, 242
European Community (EC), 41–42, 73
Executive information systems (EIS), 116
Extended Binary Coded Decimal Interchange Code (EBCDIC), 167
Extended terminal interface (ETI), 197

Facilities management, 39
Fax industry, 147–148
Fibre Distributed Data Interface (FDDI), 107
File-handling, 108
File servers, system administration, 241
File transfer access and management (FTAM), 108–110
FIPS, *See* US Federal Information Processing Standards
Flack, David, 147–148
Forms and Menu Language Interpreter (FMLI), 197–198
FORTRAN, 31, 53, 68, 69, 124, 176, 182, 184, 207
FORTRAN-77, 124
Fourth-generation languages (4GLs), 32, 198
  evolution, 182–186
  standards status, 184–185
Freedom of choice, 18
FTAM, *See* File transfer access and management
Functional standards, 112–114

Gartner Group, 3
Gates, Bill, 10, 86
General Services Administration (GSA), 128
Glyphs, 220
GOLD 400, 159–164
GOSIP, 113
Government procurements, 14
Government purchasers, incompatibility issues concerning, 16
Graphical interface, 30
Graphical user interfaces (GUI), 66, 199-211
Graphics services, 71–72, 187
Groupware, 27

Handler, Gary, 154
Hardware lock-ins, 170
Hardware manufacturers, 12
  incompatibility issues, 17–18
Hardware peripherals, standards of, 72–73
Hardware resellers, incompatibility problems concerning, 13–14

Hardware standardization, 32, 33
Hardware systems, variation in, 23
Harmonization, 113
Hayes Microcomputer Products, 47
Hewlett-Packard, 67, 220
Hi-fi industry, 8, 34
Human-computer interface (HCI), 138, 147

IBM, 67, 73, 105–106, 171, 192, 201, 236, 239, 245
IBM PC, 6, 23, 33, 34, 49, 75, 86, 89, 90, 157, 167, 183, 188, 191, 192, 210, 220
IEEE, 53, 57, 58, 60–65, 67, 68, 77, 78, 103, 107, 144, 185, 206, 224, 225, 248
Incompatibility, *See* Compatibility
Independent software vendor (ISV) advisory councils, 130
Industry standards bodies, 45
Information Builders, 183
Information Engineering Directorate (IED), 158–164
Information management, 22
Information Management System (IMS), 171
Information technology, integration with business strategy, 22–23
Information Technology Requirements Council (ITRC), 145
Informix Software, 178
Ingres, 170
Innovation, 5, 16, 34
Innovative:
  products, 47
  research, 40
Integrated network management system, 38
Integrated service digital network (ISDN), 96, 117–118
Integration devices, 21
Integration of open and proprietary hardware and software, 37
Intel 80286, 35
Intel 80386, 35
Intel 80486, 35
Intellectual property, 48
Interconnection:
  de facto standards, 104–108
  requirements, 13
  standards, 29, 35, 96–120
Inter-Departmental Electronic Mail (IDEM), 159
International market, 41
International Standards Organization (ISD), 52, 53, 54, 62, 65, 66, 68, 74, 103, 115, 177, 191, 218–220, 224, 225
Internationalization, 40, 213–226
  areas affected by, 218–221
  standards, 40, 41, 73, 224–225
  standards model, 219
  steps towards, 215–218
  user view of, 214–215
Interoperability, 27, 29, 118–119
Interoperation, 98
Interworking, 29, 98

Inventory control, 12
Investments, 3, 16
ISAM, 178–181
  and relational database compared, 179–181
  problems for existing applications, 181
  standardization status, 178–179
ISDN, *See* Integrated service digital network
ISO, *See* International Standards Organization

Jackson Method, 185
Japan, 3, 42
Japanese Institute for Standards (JIS), 52
Japanese Victor Company (JVC), 48
Job transfer and manipulation (JTM), 104, 112
Joint Academic Network (JANET), 159–164
Just-in-time (JIT) manufacturing, 115

Kerberos, 247
Kernel Group, 127
Know-how
  investment, 3, 4
  transfer, 2

Labelled security, 231
Language issues, 42
Libraries, 188, 193–196, 198
Licensing arrangements, 30
Lisa, 200
Local area networks (LANs), 106
  standards, 103
  system administration, 242
Localization requirement, 40
Low-level interface, 189

Mach, 72
Macintosh 'desktop', 201
MacPaint, 200
MacWrite, 200
Mainframe computers, system administration of, 240
Manufacturers' Automation Protocol (MAP), 113
Mapper, 184
Market entry, 41
Market forces for standards, 9–18
Market research, 8
Market size, 41
Marketing strategies, 8
Massachusetts Institute of Technology (MIT), 67
Message handling systems, 110–111
Messages, 225
Microprocessor, standardization on, 34
Microsoft, 201
Migration:
  costs, 21
  experiences, 155–166
  problems, 20–21
  tools, 21

Military applications, 234
MIT, *See* Massachusetts Institute of Technology
Mixed-vendor and mixed-technology networks, 38
Modems, 47
Modular programming techniques, 216
Morris, Geoffrey, 132, 140
Motif, 67, 125, 147, 201, 202, 210, 211
MS-DOS, 19, 20, 34, 48–49, 68, 69, 75, 92, 175, 178, 197, 222, 223, 232
Multi-national companies, 215
Multi-processing, 72
Multi-user computers, system administration, 241

National Institute of Standards and Technology (NIST), 39, 58, 128, 248
NCSC, *See* US National Computer Security Commission
Network-aware application, 28
Network Computing Architecture (NCA), 119
Network Computing Forum (NCF), 119
Network-extensible Window System (NeWS), 72, 202, 208–210
Network File System (NFS), 51, 71, 238
Network-intrinsic application, 28
Networks, 71, 98, 188
  management, 114
  management tools, 38
  Orange Book, 233–235
  system administration, 243–245
  See also Local area networks (LANs); Wide area networks (WANs)
'New Opportunities in Standards' Committee', 61
New Standards Project Committee, 61
New technology, 5
NIST, *See* National Institute of Standards and Technology
Numeric formats, 222–223, 225

Office document architecture (ODA), 104, 115
Open DeskTop (ODT), 92–94
Open distributed processing (ODP), 119
Open frontiers, 42
Open Look, 66, 125, 147, 190, 202, 210, 225
Open process, 53–54
  consensus, 54
Open Software Foundation (OSF), 67, 82, 83, 210, 233
Open systems:
  as standards, 6
  benefits of, 19
  communications sense, 6
Open systems, definition of, 5–6, 29
Open systems:
  further requirements, 37–41
  political definition, 6
  portability definition, 6
  See also Users of open systems
Open Systems Directive, The, 139–140
Open Systems Initiative, 146
Operating systems:
  interface standards, 76
  standardization, 33
  standards, 74–79
  trends in, 92
Oracle, 10, 170
Orange Book, 70–71, 228–237
OS/2, 13, 69, 94, 178, 188, 197, 222, 223, 232
OSF/1, 85, 246
OSF/MOTIF, 85
OSI, 52, 57, 66, 71, 96–120, 229
  application layers, 108–114
  functional standards, 112–114
  seven-layer model, 98–104
  standards, options in, 99–100
OSITOP, 66
Overruns, 1–2
Ozvath, John, 144

Packet switching network, 109
PAD, *See* Packet assembler/disassembler
Parties of interest, 62
Pascal, 69, 124, 176
Patents, 48
People portability, 15–16
Performance and standards, 19
Personal computers (PCs), 34
  networked, 242
  system administration, 241–242
Personalized computing, 139
PL/1, 69, 176, 182
Portability, 26, 28
  across international frontiers, 42
  definition, 6
  of people and applications, 41
  software, 6
  source code, 34
  standards, 29, 65–95
Porting costs:
  direct, 10–11
  indirect, 11–12
Porting time, 11
POSIX, 52, 55–58, 60–66, 69–72, 74, 114, 119, 138, 143, 186, 222–225, 229, 231, 236, 239, 248, 249
  history, 80
  products for implementation, 79–85
  standards, 77–79
PostScript, 71, 208–210, 220–221, 225
PostScript page description language, 48
Presentation Manager, 201, 206, 210, 211
Pricing implications, 90–91
Print control, 239
Profiles, 113
Programming languages, standards, 68
Programming tools and utilities, 68
Project Athena, 205, 247
Promotional Council for OSI (POSI), 144
Proprietary architectures, 89

Proprietary lock-in, 9, 32, 192
Proprietary systems, 14
Proprietary technology, 8, 19
Protective mechanisms, abuse of, 21
Public standards, *See* Standards
Public standards groups, *See* Standards
Purchasers, *See* Users

Rapporteur groups, 224
Regular expression matching, 221
Relational database manager, 179–180
Remote File Sharing (RFS), 71
Remote procedure calls (RPC), 119
Request for technology (RFT), 83, 85, 210
Requests for comments (RFCs), 51
Requirements for Trusted Systems, 39
Resistance to change, 4–5

Santa Cruz Operation (SCO), 75, 92
Scalable Processor Architecture (SPARC), 87
Scalability, 26–27, 28
  standards, 65–95
SCOMP, 231
Security, 21–22, 70–71, 169, 172, 227–237
  current standards, 228–35
  domains, 231
  requirement, 39
  standards, 236
  X/Open, 125
Shared technology, 19
Shearson Lehman Hutton (SLH), 153–155
Sherr, David M., 156
Simple Network Management Protocol (SNMP), 247
Single computer application, 28
Skills shortages, 1, 2, 3, 5, 14–15
SNMP, *See* Simple Network Management Protocol
Society for Worldwide Inter-bank Funds Transfer (SWIFT), 227–228
Software AG, 183
Software application, standardization of, 30–31
Software design, 41
Software developers, incompatibility problems concerning, 9–12
Software development, 41
Software distributors, incompatibility problems concerning, 12–13
Software languages, standardization of, 31–32
Software lock-ins, 170
Software Partners Program, 130
Software piracy, 245
Software tools, standardization of, 32–33
Sorting
  of character-based data, 221
  of programs, 221
Source code, 34
  portability, 34
SPAG, *See* Standards Promotion and Application Group, 66, 144
SPARC, *See* Scalable Processor Architecture
SQL, 32, 55, 58, 69, 125, 174–177, 179, 181, 218
Stacks, 113
Standardization:
  levels of, 29
  user input, 143–147
Standards, 5
  agreed, 57–58
  application environment, 66–74
  attempts to define, 7
  case for, 18
  character terminals, 191–192
  classes of, 54–56
  comment from participant, 79
  compromise, 54, 55
  cooperative, 57
  counter-arguments, 18–23
  data, 167
  data handling, 171
  de facto, 46–52, 56, 58, 66, 71, 80, 98, 104–108, 137, 178, 191, 192, 197, 198, 202, 206, 225
  delays in appearance, 57
  Electronic data interchange (EDI), 115–117
  evolution, 57
  functional, 112–114
  hardware peripherals, 72–73
  implementation possibilities, 59
  incompleteness of, 20
  industry bodies influencing, 45
  interconnection, 96–120
  internationalization, 20, 73, 224–225
  language generations, 184
  local area network (LAN), 103
  making improved, 57
  market forces, 9–18
  maximal, 54, 56
  minimal, 25, 54
  mixed networks of computers, 25
  need for, 8–9
  open systems, 6, 25
  operating system, 74–79
  operating system interface, 76
  options in, 58–60
  organizations influencing evolution of, 44
  OSI options in, 99–100
  overlapping, 56
  portability, 65–95
  POSIX, 77–79
  programming languages, 68
  public (*de jure*), 46, 48, 51, 52–53, 98
  public groups influencing, 44–45
  scalability, 65–95
  security, 236
  setting, 44–64
  several products to program, 49

speculative development, 50
system administration, 239, 245–249
technical areas and functional requirements to be covered, 25
user bodies influencing, 45–46
who needs them?, 8–24
wide area network (WAN), 103
Standards Activities Board, 61
Standards Promotion and Application Group (SPAG), 66, 144
Standards Review Committee, 61, 62
Strategic issues, 5
String searches, 221–222, 225
Structured protection, 231
Structured Systems Analysis and Design Method (SSADM), 185
Style guide, 189–190
Sun Microsystems, 51, 67, 158
Support materials, 41
SVID, *See* System V Interface Definition
SWIFT, *See* Society for Worldwide Inter-bank Funds Transfer
System administration, 238–249
academic research, 247
commercial implementations, 245
databases, 241
emerging issues, 243–245
file servers, 241
job definition, 38
local area networks (LANs), 242
location of control points, 243
mainframe computers, 240
multi-user computers, 241
networks, 243–245
personal computers (PCs), 241–242
problems of, 239–243
standards, 239, 245–249
tools, 38
user view, 238–239
wide-area networks (WANs), 243
System Application Architecture (SAA), 106
System Network Architecture (SNA), 35–36, 105–106
System upgrades, 26
Systems houses, 13
System V Interface Definition (SVID), 56, 58, 82, 195, 196, 198, 249

Taiwan, 3
TCP/IP, *See* Transmission Control Protocol/Internet Protocol
Teather, Bill, 152
Technical Office Protocol (TOP), 113
Technical support on application products, 12
Termcap, 193–195
Terminfo, 195
Testing of claimed conformance, 34–90
Time delays, 12
Timeliness, 172
Token Bus, 107
Token Ring, 103, 106, 107, 242
Training and retraining, 3
Training programs, 37
Transaction processing systems, 70
Transmission Control Protocol/Internet Protocol (TCP/IP), 51, 71, 104–105, 247
Transport layer interface (TLI), 99
Trial-Use document, 63
Trusted computer systems, 228–237
levels of trust, 229–231
Trusted Computer Systems Evaluation Criteria (TCSEC) (*See* Orange Book)
Trusted Network Interpretation (TNI), 235

UKGOSIP, 113
Uniforum, 76–78, 143–144, 224
United States, 41
UNIVAC, 157
Universities, 14
UNIX, 11, 13, 19, 20, 23, 33, 36, 50, 56, 65, 66, 69, 70, 72–76, 80, 86, 92, 124, 138, 143, 144, 149–151, 155, 174, 178, 183, 185, 192, 196, 197, 206, 208, 214, 222, 224, 227, 228, 238
and Orange Book, 231–233
de facto terminal handling standards in, 193–197
fragmented market, 86
in non-UNIX environments, 155–158
operating system, 49
Unix International (UI), 82
UNIX Software Operation (USO), 82
UNIX System V, 81–82, 209, 225, 239, 245–246
US Federal Information Processing Standards (FIPS), 39, 58, 231, 248
US National Computer Security Commission (NCSC), 228, 235, 236
USENET, 36
User bodies influencing standards, 45–46
User Environment Component (UEC), 85
User-friendly technology, 30
User interface, 66–67, 187–212
character-based, 191–198
model and application generator, 198
multi-layer model, 188
standardization, 29–30
technical overview, 187–91
toolkit, 189
User needs, 172–174
User prompts, 223
User responses, 223
Users of open systems, 142–166
advice to, 164–166
case histories, 148–155
definition, 142
input to standardization processes, 143–147
USGOSIP, 118

Value-added resellers (VARs), 12, 13
Venture capital, 16
Verified design, 231
VHS, 9, 49
Video cassettes, 48
Video industry, 9
Virtual terminal standards (VT), 112
VMS, 69, 178, 197
VT100, 191

Wastage, 3
Wide area networks (WANs), 107–108
  standards, 103
  system administration, 243
Widgets, 207
Wild card characters, 221
WIMPs, 200
Window system, 189
Window systems, 202–211
Word-processing package, 30–31
WordPerfect, 30–31
Working groups, 54, 61, 62
Workstation computers, 242

X3, 108
X.25, 103, 108
X28, 108
X29, 108
X.75, 108
X.400, 110–111, 158–164
X.500, 111–112
XENIX, 75, 91–92
Xenix, binary compatibility, 91
Xerox's Palo Alto Research Center (PARC), 200
Xionics Message Transfer Agent (MTA), 162–164
Xlib, 207
X/Open, 42, 74, 94, 100, 121–141, 224, 225
  call to action, 140
  Common Applications Environment (CAE), 122–6, 153–155
  completion of cycle, 129–130
  contributors to, 121
  data management, 125
  history, 121
  market maturity issue, 137
  membership of, 121
  mission, 122–123, 135, 139
  networking, 125
  operating system services, 124
  partnership, 136
  perspective for future, 132–135
  portability guide (XPG), 124
  programming languages, 124–125
  Requirements Conference, 146–147
  requirements process, 145–146
  security, 125
  shareholders, 121
  Software Partners Program, 130
  technical process, 126–130
  transport interface (XTI), 125
  User Advisory Council, 153
  user and ISV advisory councils, 130
  verification and branding program, 128–129
  window management, 125
  working groups, 127–128
X/Open ISAM, 178–179
X/Open Portability Guide (XPG), 126, 129, 135, 177, 178–179, 182, 195, 196, 224, 225, 236, 239, 249
X/Open Requirements Conference (XTRA), 132
X/Open transport interface (XTI), 99
XPG3, 125
X terminals, 67
XTI, 100
XTRA process, 139–140
X Window Recommended Practice, 206
X Window System, 51, 67, 125, 202, 205–207, 210, 211, 220–221, 225, 247
X Window Toolkit Application Programming Interface, 206